Hanns Hatt • Regine Dee

The Power of Scent

How Smells Seduce and Heal

Hanns Hatt
Ruhr-Universität Bochum
Bochum, Germany

Regine Dee
Hamburg, Germany

ISBN 978-3-031-88523-5   ISBN 978-3-031-88524-2 (eBook)
https://doi.org/10.1007/978-3-031-88524-2

This book is a translation of the original German edition "Die Lust am Duft" by Hanns Hatt and Regine Dee, published by Springer-Verlag GmbH, DE in 2023. The translation was done with the help of an artificial intelligence machine translation tool. The authors subsequently revised and added new content, and the text was revised by a professional copy-editor to enhance its style. Springer Nature works continuously to further the development of tools for the production of books and on the related technologies to support the authors.

© The Editor(s) (if applicable) and The Author(s), under exclusive license to Springer Nature Switzerland AG 2025

"Die Lust am Duft" by Hanns Hatt and Regine Dee, published by Springer-Verlag GmbH 2023

This work is subject to copyright. All rights are solely and exclusively licensed by the Publisher, whether the whole or part of the material is concerned, specifically the rights of reprinting, reuse of illustrations, recitation, broadcasting, reproduction on microfilms or in any other physical way, and transmission or information storage and retrieval, electronic adaptation, computer software, or by similar or dissimilar methodology now known or hereafter developed.

The use of general descriptive names, registered names, trademarks, service marks, etc. in this publication does not imply, even in the absence of a specific statement, that such names are exempt from the relevant protective laws and regulations and therefore free for general use.

The publisher, the authors and the editors are safe to assume that the advice and information in this book are believed to be true and accurate at the date of publication. Neither the publisher nor the authors or the editors give a warranty, expressed or implied, with respect to the material contained herein or for any errors or omissions that may have been made. The publisher remains neutral with regard to jurisdictional claims in published maps and institutional affiliations.

Cover image created by AI from Adobe Stock (807207085), modified by Deblik

This Springer imprint is published by the registered company Springer Nature Switzerland AG
The registered company address is: Gewerbestrasse 11, 6330 Cham, Switzerland

If disposing of this product, please recycle the paper.

The Power of Scent

The Power of Scent

Nothing works without the nose. We need it for every breath. Every day, it warms up 10,000 liters of breath to a body-friendly 34 degrees, moistens it with nasal mucus, and simultaneously filters out the dirt and transports it away. It enables us to smell and taste, and it warns us of fire, toxic substances, and other dangers. It decides which people we find seductive or disgusting. We turn up our noses when we don't like the scent of another person and decide "I can't stand them." Often, we don't even know why, because the nose decides independent of our intellect. This would seem to contradict our self-image: Do we not have control over our own lives?

In fact, fragrances can do more than even that which we have long suspected. Scientists have shown that many organs, especially the skin and the lungs, absorb fragrances and react to them. And the latest results of research are even more exciting: As we have been able to prove, tumor cells similarly react when certain fragrances affect them. They stop growing, and some even disappear. This represents hope for many sick people, because it is possible that entirely new paths for cancer therapy may emerge. Conversely, we now also know that tumor cells themselves emit odors. Trained dogs can already recognize today whether a person has, for example, a bladder or lung tumor. Perhaps there will soon be biosensors that work like dog noses, making elaborate tests unnecessary.

For a long time, the nose was neglected, ridiculed, or even despised. Snobbishly, humans looked down on the animals that sniffed the ground. Wasn't there also something animalistic about the nose? Messages driven by smell from sweat and other body fluids stimulated the imagination. The church saw religious zeal threatened by seductive scents and feared sexual excesses. Philosophers derisively dismissed smelling as a lower sense, even our most unnecessary, considering it a sense of pleasure, not intellect.

Smelling is also considered one of the lower senses in physiology, together with tasting and touching. But the truth is: The nose is our most sensitive sensory organ and deeply affects our lives.

It reminds us of distant days of childhood, when happiness began, when grandmother baked a cake. The smell of a freshly baked apple pie will forever be associated with a feeling of satisfaction and joy. Out of the blue, we then feel happy, without knowing why. Smelling, this ancient sense essential for both survival and enjoyment, has thus led us into the past, because the nose sends all scent signals straight into the memory and emotion center of the brain, which suddenly takes us back to distant childhood days, without us thinking much about it or being able to resist.

We particularly love the scents we know. Both the familiar family smell and the smell of home are something very special; they stay with us for a lifetime and always evoke the same feelings as when we first got to know them. Anyone who grew up on a farm likes to walk where it smells of cow dung. Anyone who, as a child, went on vacation to the sea loves the salty scent of the waves.

Scents and scent preferences are very individual. Everyone experiences smells differently. Everyone also sends out their very own scent messages, which we don't even know ourselves. We may notice our sweat smell and wash ourselves, but this does not eliminate all scent messages, a disturbing notion. We usually are only used to accepting instinctive messages from our dog! Even when walking with the beloved pet, we suspect that there is another whole world of smells. An invisible world, completely closed off to humans: the scent language of animals. When the dog pees on every tree, he leaves important information for all subsequent species: "I was here 10 minutes ago, I am a young, strong dachshund and can easily handle anything I need to. Watch out!" Each species has its own chemical language, its individual pheromones. Through these, others are warned of dangers and informed about important everyday things such as age, strength, readiness to mate, and where they find their food sources. And—very importantly—the chemical language ensures the right sex partner!

In extensive experiments, researchers have searched for answers—most often in the armpits of men—to the exciting question: Can it be that humans also possess these invisible powers of seduction? Do humans also send out pheromone messages? Indeed, the researchers found: When women smell men's sweat, areas of the brain are activated that are associated with sexuality and partner choice. Apparently, they react to the individual male scent in the search for the best father for their children. And, in doing so, they do not decide based on beauty or intelligence, but rather exclusively based on gene pool. Men, by the way, don't care much about the women's genetic makeup. Other seduction techniques work on them.

Perfumes have been around since ancient times and have never gone out of fashion. Thousands of years ago, the fragrances of frankincense and cedar were meant to appease the gods. The Egyptians embalmed their dead with fragrant essences, and Cleopatra successfully seduced the Roman general Marc Antony with rose petals and jasmine. Through the distillation art of the Arabs, a flourishing trade emerged that quickly reached Italy and France. The center of sea trade was then the Republic of Venice. Catherine de Medici, through her marriage, exported the art of perfume to France, where it was very welcome. Water had just been exposed as the cause of all kinds of diseases. Perfume came in handy to mitigate the worst smell disasters. Eau de Toilette was invented. The perfume consumption of the royal mistress Madame de Pompadour is legendary. Probably only surpassed by that of Napoleon, who, according to tradition, poured a liter of the popular Eau de Cologne 4711 over himself every day. To the flowers, herbs, and resins used were later added synthetic scents, the famous perfume No.5 by Coco Chanel from 1921 being the first example of this.

Many scents and oils are suitable as remedies. Here, too, there is a long tradition, from Hildegard of Bingen to modern aromatherapy. Inhaled or rubbed in, oils and essences can have amazing effects. Today, we know why this is. As our research at the Ruhr University Bochum has proven, olfactory receptors not only exist in the nose but also in many other organs. This opens up a multitude of possibilities for therapy. The aforementioned fact that tumor cells react to certain scents is, of course, particularly exciting. Now, these findings need to be deepened and made suitable for practice. The world of scents remains exciting and will continue to surprise us in the future.

Bochum, Germany	Hanns Hatt
Hamburg, Germany	Regine Dee

Contents

1	**How Smelling Works**	1
	Lightning-Fast Reaction in Danger	2
	Smelling Needs to Be Learned	3
	The Fragrance Alphabet Has 400 Letters	4
2	**The Nose Never Sleeps**	5
	When the Nose Switches Off	6
	Right and Left Noses	7
3	**Three Specialists for Perfect Taste**	9
	Spinach? Spit it Out!	10
	The Chocolate Nerve	10
4	**Happiness Makers for the Brain**	13
	Calories for a Satisfied Stomach	14
	Some Like it Hot	15
5	**The Good Sense of Smell in Animals**	17
	Animal Super Noses	18
	Practice Makes Perfect	19
6	**The Secret Messages of Pheromones**	21
	No Sex Without Scent	22
	Conspecifics Are Warned	23

7	**The Scent of Flying Sex Machines**	25
	The Language of Pheromones	26
	Specialists at Work	26
8	**Deception and Trickery with Irresistible Scents**	29
	Scenting by the Clock	30
	No Sex, Please!	30
9	**Nobody Smells as Good as You**	33
	Tempting Messages	34
	The Decisive Body Odor	34
	The Best Perfume for Women	35
10	**Cold Sweat and Baby Scent**	37
	The Smell of Trust and Friendship	38
11	**Diagnostics with the Nose**	41
	When Sweat Smells Like Urine	42
	Dogs and Rats as Lifesavers	43
	Super Nose Discovers Parkinson's	44
12	**Smelling with Skin and Hair**	45
	Hair Roots Recognize Sandalwood Scent	46
13	**Scents as Therapy Aids**	49
	Herbal Scents for Digestion	50
	Fragrances Against Respiratory Diseases	50
14	**Fragrances Against Tumor Cells**	53
	New Approaches for Diagnosis and Treatment	54
15	**When Scents Go to Our Heads**	55
	GABA Receptors for Restful Sleep	56
	Tonic Water Against Leg Cramps	57

16	**When the Nose Goes Blind**	59
	Fresh Olfactory Cells Every Month	60
	When Orange Juice Smells Like Gasoline	61
	The Nose Ages Too	62
17	**Of Stinky Fruits and Moldy Cheese**	63
	Family Scent Connects	64
	Popular Worldwide: Oranges	65
18	**Get Slim Faster with Bitter Substances**	67
	Bitterness Drives Away the Addiction to Sweetness	68
	Super Tasters Get Full Sooner	68
19	**The Scent of Christmas**	71
	Cookies Instead of Pills	72
	The Secret of Gingerbread	73
	Vanilla Against Christmas Stress	74
20	**The Sophisticated Palate Games of Wine**	75
	Noble Barrels and Oak Chips	76
	Valuable Cork	77
	The Dose Decides	77
21	**Whether Drugs or Truffles: Dogs are Perfect Sniffers**	79
	On the Trail of Rare Species	80
	Dogs as Truffle Hunters	80
22	**Detection Dogs in Medical Use**	83
23	**Animal Super Noses Help Humans**	85
24	**Divine Fragrances and Worldly Perfumes**	89
	Romance with Rose Petals	90
	World Success from Glockengasse	91

25	**In the Frenzy of Fragrances**	93
	The Personal Scent Decides	94
	Effective Scents for Relaxation and Trust	95
26	**Marketing with a Feel-Good Factor**	97
	Scent Landscapes for Women's Fashion	98
	The Special Scent of a New Car	99
	Corporate Scent: A Touch of Well-Being	100
27	**From Smell Training to Brain Jogging**	101
	Therapies with Varying Success	102
	Smell Exercises for Greater Quality of Life	102
	Smell Training as Brain Jogging	103
28	**The Future: eNoses in Medicine**	105
	Detecting Diseases and Germs Earlier	106
29	**Truffles, Tea, and Bugs: Searching for Species with the eNose**	109
	NOSI Warns of Rotten Fruit—and Bedbugs	110
30	**Nanoprostheses and Natural Replacement Noses**	113
	E-Noses from Natural Odor Receptors	114

1

How Smelling Works

> Smelling is a complex interplay of scent molecules, olfactory cells, and our brain. We begin training our sense of smell even before birth, as the nose is supposed to protect us from dangers and guide us safely through life. The fact that it also awakens love, desire, and well-being is a wonderful feature of nature.

The nose is our most sensitive sensory organ and deeply influences our lives—mostly unconsciously. Only a few molecules are enough and we revel in the scents of spring, a perfume, or a suddenly very interesting person. The nose explores all the aromas of this world for us, from the finest red wine to the most exquisite truffle, while simultaneously awakening memories of long past childhood or beautiful vacation days. Mom's strawberry jam—heavenly! The herbal scent of an alpine meadow—like the carefree summer holidays of school time!

20 million olfactory cells, 10 million per nostril on the surface of a euro coin, are not only out for pleasure, but also pick up the faintest traces when danger threatens. The forest is on fire? The nose warns us in advance. The fish smells rotten? Poison lurks here. Someone spreading foul smells? Beware of disease! While eyes and ears are directed into the distance, the nose and mouth represent the last line of defense to protect humans. Spit it out, avoid contact, hold your breath and then get away as quickly as possible, is the urgent advice. Unlike light stimuli and sounds, scent information is long-lasting and spreads over large distances. In such situations, the nose can help us survive—or smell who is already in the office in the elevator in the morning.

So how does the aftershave scent of a colleague get into our nose and the information about it into our olfactory brain? The nose consists of three levels. The top one contains 20 million olfactory cells. They consist of an oval cell body that has a small thickening at one end. From this, about 20–30 finger-like extensions (cilia) protrude into the nasal mucus. A centimeter-long thin nerve fiber (axon) grows towards the brain from the olfactory cell body. It transmits information from the nose to the olfactory brain. But this only works because the skull bone has holes like a sieve here, which is why this part is logically called the ethmoid bone.

Lightning-Fast Reaction in Danger

In humans, 400 different types of olfactory cells, equipped with scent sensors, so-called scent receptors, have been identified. Each cell is specialized in a group of scent molecules, such as vanillin, musk, or butyric acid, as it only produces one of the 400 types of receptors. This covers our entire world of smells. If a vanillin molecule is in the breath, it can activate its corresponding receptor like a key being inserted into the right lock. The olfactory cell amplifies the chemical signal and converts it into an electrical pulse, which reaches the olfactory bulb in the brain via a thin nerve thread.

What is fascinating is that all olfactory cells of a certain type (approx. 50,000) end in the same brain cell with their nerve extensions in the olfactory brain. So in the human olfactory brain, there are 400 small cell clusters (glomeruli) that correspond to the 400 types of olfactory cells. All glomeruli together form the olfactory bulb. A magnificent system that works lightning fast. And even faster in danger! This has now been discovered by researchers from the Swedish Karolinska Institute. In their study, they were able to demonstrate that the cells of the olfactory bulb react to smells within 50 to a few hundred milliseconds. For smells that the brain perceives as are classified as unpleasant or potentially dangerous, the reaction occurs up to 15 times faster than with positively rated smells. What is particularly interesting is that the motor system of the test subjects was immediately activated by the unpleasant smells: they involuntarily leaned back slightly. The researchers concluded from the results that an evaluation of the smell is already made in the olfactory bulb, before the human perceives this smell consciously as pleasant or unpleasant. They assume that previous smell experiences and biological imprints play a significant role in this.

Smelling Needs to Be Learned

Even the mucus, which often seems annoying to us and paralyzes all olfactory cells when we have a cold, is in truth a misunderstood genius. Each nasal cavity is—apart from the entrance—equipped with mucous membrane. This is supposed to warm and moisten the inhaled air and also serves to clean the air. The mucus itself consists of a special protein composition. It not only protects the cells, but is even actively involved in the transport of fragrances to the sensory cells. Fragrance molecules are usually hardly water-soluble, but should nevertheless reach the olfactory cells through the watery mucus. That's why nature has developed binding proteins that transport the fragrances through the nasal mucus to the olfactory cell. Nature was not stingy in this regard: over 100 different proteins (this applies to the mouse, humans only have 10) bind the fragrances, with each protein specialized in transporting a certain group of fragrances to the olfactory cells.

But before everything works smoothly, our olfactory system must first learn. The fragrance school begins already in the womb. The researchers knew this from experiments with rabbits. They gave pregnant rabbits juniper to eat, which is not actually one of their favorite dishes. But lo and behold: the rabbit babies of these mothers later all preferred to nibble on juniper twigs instead of the otherwise so popular herbs and dandelion leaves. And what happens with human embryos? Researchers asked two groups of expectant mothers in the last third of pregnancy and during breastfeeding to either only drink water or, in the second group, also to consume carrot juice. The result was clear: The babies who were already familiar with the carrot taste made fewer faces and ate their portion of carrot puree faster. Because human babies also learn through the amniotic fluid early on what their mothers like to eat or don't like at all. So it's not surprising that mothers who love garlic also have children who love garlic. And, as scientists have shown, the same happened with pickles and also chocolate: these preferences of the pregnant women were also transferred to their babies. From the 26th week of pregnancy, the olfactory cells and their connections into the brain are already fully developed. Thus, the information from the nose can be collected and processed in the olfactory bulb. They are then transmitted via two thick nerve strands directly into the memory center (hippocampus) and into the emotional center (limbic system) and linked together.

Other fragrance experiences of the mother, especially when they are associated with strong emotions, are also smelled and learned. Thus the embryo takes on the positive and negative emotions of the mother and stores them.

Therefore, a baby is already born with fragrance preferences and dislikes and can still remember a fragrance years after birth that it only knows from the time in the womb. This is a very special achievement, because the embryo already has to store complex fragrance patterns and learn them.

The Fragrance Alphabet Has 400 Letters

Because most smells consist not only of one type of fragrance molecules, but a whole mixture. Coffee aroma, for example, is composed of more than 200 different fragrances. They all activate "their" receptor types, so that a typical Coffee activation patterns emerge. The scent alphabet has 400 letters, corresponding to the 400 types of olfactory receptors. "Scent words" can be 100 or more letters long. No wonder that scents are much more difficult to learn than words. Over time, especially with a lot of practice, the nose learns many such "scent words" and recognizes the smells again. For some smells, humans even have a finer nose than the dog with its legendary super nose. For example, we can smell bananas very well, which are important for our nutrition. For the dog, bananas have no meaning, so they are quite indifferent to him. In general, humans smell much better than was long believed. However, we can't compete with rats or elephants, which have the best sense of smell, as studies have shown. These animals have a multitude of receptors and sensory cells and experience the whole world through their noses.

The hot wire that scent information takes into the oldest areas of the brain causes smells to trigger memories and feelings directly, without the person really knowing what is happening or even being able to make a rational decision beforehand, because only a few scent information passes through the gate to consciousness, the thalamus. Unconsciously and involuntarily we feel disgust, desire or pleasure. How we evaluate certain scents is not genetically programmed, but learned and depends on our experience and upbringing. A romantic vacation in Provence and we rave about lavender from then on. The unloved aunt from childhood, who tried to dispel her old age smell with lavender, makes us find the scent repulsive for life.

By the way, humans can do a lot to train their sense of smell. The more he practices learning and recognizing scents, the more scents and aromas he offers to the nose, the better it will function. Daily smelling exercises are best. Regardless of age, you can start at any time. Give it a try: identify the spices and herbs at your next restaurant meal, discover three unknown smells at your next shopping trip to the farmers market.

2

The Nose Never Sleeps

> Smelling is supposed to protect us day and night—a mammoth task. The air surrounding us is analyzed around the clock, 20 million olfactory cells constantly on the alert for possible dangers. Only sometimes is one side of the nose allowed to rest a bit. Which one? This varies from person to person, as people differ in being right-nosed and left-nosed.

At the end of the day, when humans sink tiredly into their pillows, they close their eyes, plug their ears against noise, and sleep. No such rest is accorded the nose. It works nonstop. Day and night, 24 h, a lifetime. While humans surrender to their dreams, their noses watch over them and continue to sniff. As long as we breathe, we smell. Completely automatic. We can't turn off the nose, because, after all, you can't hold your breath for too long. Some scents are particularly pleasing to the nose, such as the scent of oranges or roses. In that case, the human dreams something beautiful. Conversely—as our experiments in sleep labs have shown—stench can have the opposite effect: When faced with the foul smell of feces, people report unpleasant experiences in their dreams upon waking.

The scent of a beloved partner, on the other hand, has a positive effect on men's dream content. This is good news for women. The bad news: the scent of oranges produces the same effects. So, if the woman is traveling, she should leave an orange on the pillow to continue to provide the man with pleasant dreams, at least. Women often use a T-shirt worn by their beloved man or child for better sleep and nicer dreams. Rose scent is attributed another very useful property: Brain researchers found that this scent can awaken memories of things learned during the day while we sleep. They wafted scent over one group of subjects during the deep sleep phase, while another group slept in an odor-free room. The next morning, the researchers compared the results and found: The memory of what was learned was significantly better in the rose scent group. The tireless nose had therefore continued to work during sleep, supporting learning.

When the Nose Switches Off

It often happens that the nose has had enough, so to speak, being completely fed up with a certain smell. This can be an elegant cosmetic scent that has been wafting around us for a long time, as well as an unpleasant smell of garlic or sweat. Then, the nose simply switches off and protects our brain from overstimulation. Scientists speak of adaptation: The nose has become accustomed to a smell, the human no longer perceives it. Our olfactory cells no longer send electrical pulses to the brain. This affects our body odor, as well as our favorite perfume in the morning. If we therefore spray generously several times to be safe, we may reap a nose wrinkle from others, because they think we have bathed in the perfume bottle.

However, adaptation also makes many an olfactory challenge of everyday life bearable. It benefits us in crowded commuter trains at the end of the day, as well as in the gym or during a conversation with the boss who had lunch at the Greek restaurant: After some time, the stinking experience is over and the nose can only be switched on again by a breath of fresh air. Coffee aroma also clears our "scent hard drive" and makes the nose once again receptive to new things. Some employees of perfumeries know this and place small plates of coffee beans between the perfume sample bottles. A few sniffs and you can continue with the perfume purchase. To this day, there is no scientific explanation for why the scent of coffee has such an effect.

Dog noses also know adaptation. When they follow a trail, they therefore use a trick: They never stay too long in one place. They do not follow the scent cloud of the prey directly, but zigzag to the left and right of the scent trail to take a few breaths of fresh air now and then. In this way, they constantly free their noses from too many scent molecules and prevent adaptation. By the way, male moths use the same strategy: They follow the scent trail of a sex pheromone in a zigzag course to find the moth lady of their dreams among trees and bushes.

Right and Left Noses

In humans, not all 20 million olfactory cells are permanently needed for the analysis of smells. Therefore, one side of the nose is always allowed to rest a bit. People differ in this respect between right and left noses. Some smell through the right nostril 80% of the day, and through the left 20%, while, for others, it is the other way around. Only when sniffing intensively, when a scent particularly interests us, are both active. In regard to the recognition of smells, this does not matter: What you learn on the left, you also recognize on the right. Trained noses can theoretically distinguish over a billion scents in this way. A ten trillionth of a gram of mercaptan—a monoterpene from grapefruit—is enough. A drop of it distributed in Cologne Cathedral would already be noticeable.

While humans long believed that eyes and ears were much more important to our species than the nose, our body teaches us otherwise. It takes the nose very seriously and invests very carefully in its chemical early warning system. It reserves 3% of its genes solely for the production of olfactory receptors. Every month, it organizes a fresh cell cure for all of its olfactory cells to keep them fit and functional. Stem cells replace the old olfactory cells, which may be damaged by cold viruses or environmental toxins, and supply us with a completely new set of olfactory equipment every 4 weeks. An incredible achievement of the nerve tissue: 1000 new olfactory sensory cells are created every minute of our lives and their nerve extensions grow to the same brain sensory cell as their predecessors. And all this without us feeling the slightest thing. Therefore, we do not need to worry if we suddenly cannot smell or taste anything during a flu infection. This is normal and, in most cases, a temporary condition. As long as a disease does not attack the stem cells, the process runs smoothly. Only in old age does the ability to regenerate gradually decrease.

3

Three Specialists for Perfect Taste

> People talk about refined palates and culinary delights and forget the most important gourmet: the nose. Only the nose takes in all of the aromas of our food and passes them on to the brain. Together with the taste nerves of the tongue and the trigeminal facial nerve, it ensures perfect enjoyment.

Anyone who has ever had a real cold knows the phenomenon of a blocked nose: Nothing can be "tasted" anymore. The mucous membranes are swollen, no fragrance reaches the receptors, you can't tell an apple from a piece of kohlrabi. This can have advantages when it comes to foul-smelling medicine—you don't need to hold your nose anymore. But with a delicious meal, it's a real shame, because the nose plays a major role. And when we say, "That tasted great to me," we actually do not mean the taste, but a combination of smell, taste, and the sensation of the trigeminal nerve.

The tongue is responsible for the actual tasting. In contrast to the sensitive nose, which perceives thousands of delicate flower and spice aromas, the tongue is rather simply structured. It has taste buds that make it look like a peeled navel orange, but its four receptor types in the sensory cells (orange slices) only pick up the basics: sugar, salt, acid, and bitter substances. The taste substances trigger an excitation in the taste buds, which is received by the taste nerve and transmitted as electrical impulses to the brain. There, they end up in the sections for emotions, but also in the sensorimotor fields for temperature, pain, and facial expressions—which is why you inevitably make a face when sucking on lemons. They also migrate to higher brain structures, where the scent stimuli from the nose also arrive.

Spinach? Spit it Out!

The sour taste protects us from too much acid, because this taste is actually a warning signal of a low pH value, which occurs in, among other things, unripe fruits and spoiled food. We are best equipped for tasting bitter substances. There are 25 different receptor types responsible for this, with only three for sweet and one each for salty and sour. The proportion of the different types changes over the course of life. Children (up to 7 years) have few sweet receptors, so they can't get enough sweetness, but they have more receptors for bitter substances than adults. That's why they don't like bitter vegetables and refuse Brussels sprouts and spinach. Not to annoy their mothers, but to protect themselves from bitter things like nicotine, coffee, and beer, which are harmful to them. Plant toxins are often bitter too. And while sweet, salty, and sour are perceived everywhere on the tongue, the bitter receptors are strategically located at the very back of the tongue, near the vomiting center. They are the body's last analysis station before swallowing. Or spitting out. Tough luck for mom!

Chewing brings the nose into play. It is inseparably connected to the mouth. Through a special tube, the released aromas travel, so to speak, from the back to the upper floor, where the olfactory cells are located. This process is called "retro nasal smelling". 400 different scent receptors begin to analyze the incoming scent mixture. Not an easy task, because food contains a mixture of different fragrances. Learning these scent patterns is a complicated and laborious process, and it is especially important for children to smell and taste natural aromas first—the original product—before they strain their noses with imitations from artificial flavors. Gourmets and sommeliers have trained their noses for years and can recognize foods or wines that differ by only a few components among the hundreds of fragrances.

So, there are no "refined palates" at all. And even with the "gastronomic delights", it's a bit tricky. The palate can't taste anything, but it can feel: the cool fizz of champagne, the creamy consistency of mousse au chocolate, or the crunchiness when chewing cornflakes. The trigeminal nerve is responsible for this haptic pleasure, the "mouth feeling", the third in the trio of chemosensory systems.

The Chocolate Nerve

It is actually a warning and pain nerve that can be activated anywhere in the mouth and nose. Its temperature receptors report cold when we suck on menthol, lukewarm temperatures when thymol from the oil of thyme or oregano is nearby, or heat from really spicy curry sauce. We break out in sweat and even feel pain when the cook has stirred too much capsaicin in the form of pepper into the sauce. But what hurts in excess adds the spice to the meal in the right dose.

The trigeminal nerve should not be underestimated as a sensory nerve either. It creates the pleasant feeling of cream and fat in the mouth and lets us feel the delicate melt of chocolate. It also ensures that the interplay of different textures and flavors makes us happy and satisfied. The reverse is also true: Only when we are really feeling well can we enjoy food. This was determined by British scientists in an experiment. Those who produce more serotonin when happy can perceive sweetness better. An increase in the level of noradrenaline made the subjects more sensitive to bitter and sour. In people with depression or anxiety disorders, the production of both substances is reduced.

The trigeminal nerve not only benefits the cake and chocolate eater, but also the wine drinker. Our laboratory at the Ruhr University Bochum was able to show that the barrique sensation is exclusively mediated by receptors on the trigeminal nerve. This fuzzy-rough mouthfeel, the so-called astringency, is triggered by chemical substances such as tannins. When a wine is stored in an oak barrel, the typical barrique taste is created. However, we have now been able to recreate this taste in the laboratory and "re-barrique" the wine as desired. One drop corresponds to one year of storage in an oak barrel.

Whether fabricated or real: In the reward and happiness center of our brain, the nucleus accumbens, all information from the senses of smell, taste, and tactility are linked. And when everything runs perfectly, we are rewarded with the optimal feeling of happiness of a completely perfect meal.

4

Happiness Makers for the Brain

> We melt at the sight of chocolate, and even a hint of vanilla pudding makes our mouths water. Why? Because they activate our brain's reward system. Just like some other foods that seduce us with happiness hormones.

Chocolate is the favorite treat of the Germans, and the absolute star among the happiness makers. It consists of 600 flavor components. Studies from the USA show that 50% of Americans would rather give up sex than chocolate. Our brain rejoices at the sight of it, because it knows and remembers the velvety feeling on the tongue and the satisfied happiness after enjoying chocolate. The nucleus accumbens, a small nucleus in the forebrain responsible for pleasure and happiness, is particularly alert in this regard. It is part of the reward system, a whole network that can be activated in various ways. A small compliment or a hearty laugh, recognition and tenderness, or the enjoyment of a great meal is enough. Experiments show that the combination of sugar and fat is particularly effective, and chocolate is unbeatable in this respect. The miraculous nucleus is stimulated by the happiness-bringing hormone dopamine and then sends its excitation potentials to other brain areas, such as the hippocampus, the seat of memory, and the amygdala, which is responsible for emotions, pleasure, and joy.

Calories for a Satisfied Stomach

The reward system kicks in when a person eats things they like and that have a positive emotional meaning for them. This can be cheese bread and sour herring for some, but for most people, it is sweet, preferably fatty, energy-rich foods. This is because we had to ensure, during evolution, that we supplied our bodies with sufficient calories. It rewards us for this with a good mood and a thoroughly satisfied stomach.

In the case of chocolate, even a photo is enough to trigger the happiness kick. Since we like to comfort or reward ourselves with chocolate, the image in front of our eyes, and especially the smell of chocolate in our noses, is enough to trigger real fireworks of activity in the brain's comfort centers. London neuropsychologists found that the alpha and beta activity of brain waves were simultaneously increased by the smell of chocolate. Alpha waves are found in relaxed people, beta waves in attentive and excited people. The smell of chocolate seems to be able to do both: relax and keep you mentally awake at the same time. This is apparently due to the ingredients. The phenylethylamine of the cocoa bean is a real happiness drug and is also found in the blood of people newly in love. In addition, there is anandamide, a drug also found in the cannabis plant.

Bananas are as effective as chocolate. English scientists were able to observe this using magnetic resonance imaging. With the sweet smell of bananas, as with chocolate, a chain reaction starts that first stimulates the production of insulin, which, in turn, triggers the messenger substance tryptophan, from which serotonin is then formed, an important component of our happiness hormone cocktail. It tells us when we are satiated and satisfied. We are happy with bananas! And also healthy, one might add. So healthy that you can feed on their vitamins, minerals, enzymes, and hormone substances for days without suffering from deficiency symptoms, especially since bananas with a high carbohydrate content are also very calorie-rich.

Those who don't like bananas will definitely be happy with vanilla, the doping scent of our childhood. It awakens memories of the distant days when we were carefree and secure, enjoying the delicious, vanilla-scented milk of the mother's breast. Even infants smell like Vanilla, especially on the nape of the neck and at the fontanel. This scent shapes us for life and continually ensures an adequate production of happiness hormones. Whether as ice cream or pudding, refined with rum or anise notes, as the French like it, or with a light touch of egg and condensed milk, as the English like it: studies have shown that vanilla calms, lifts the mood and helps protect against fear, stress and burnout symptoms. Sometimes you don't even notice that vanilla is involved, because the aroma is also used to mask bitterness or reduce sharpness. Vanilla is also found in chocolate, in many dairy products and in ketchup. Vanillin was, by the way, the first natural aroma to be replicated in the lab. Its production in 1874 helped the manufacturers of puddings and baked goods become independent of the expensive vanilla pods from overseas.

Some Like it Hot

A fiery harbinger of happiness is the chili pepper. All pepper-related plants contain the active ingredient capsaicin, which can activate the warning and pain nerve Nervus trigeminus. And because the "capsaicin sensor" is also responsible for the perception of heat, chili peppers ("hot chili") can trigger the same reactions as hot soup. In extreme cases, it can even lead to heavy sweating and burning pain. Our mouth seems to glow, although, objectively, the temperature has not changed. The sensation arises solely in the brain. But just like with a burn, the sweat is supposed to cool us down, while the release of endorphins protects us from pain and works similarly to morphine. Together with the massive increase in blood circulation, the endorphin kick ensures that the person falls into a rewarding rush of happiness. It can even become addictive and tempt people to spice up their food to be more and more fiery.

The fifth happiness maker is cheese. Not without reason is it often the conclusion of a good meal. Cheese is rich in tryptophan, like chocolate and bananas, and, of course, in many important proteins and carbohydrates. If you drink a glass of red wine with the cheese, both intensify in the stimulation of the reward center, because red wine slows down the breakdown of serotonin. Serotonin not only makes you happy, but also more self-confident. Researchers have only recently found this out. People who ate a tryptophan-poor diet during the experiment soon began to neglect their own interests. A control group that regularly ate cheese sandwiches, on the other hand, shone with self-confidence and determination. A publication in the renowned journal "Science" puts the substance serotonin in a completely new light. Up to now, negotiation skills and clarity of thought have been attributed primarily to an optimal blood sugar level—if it is too low, the brain becomes cloudy. "Much more important is the serotonin level," the Cambridge doctors now assert, assigning serotonin a decisive role in controlling our emotions and our social behavior.

5

The Good Sense of Smell in Animals

> The nose plays a central role in the animal kingdom. It detects dangers and finds both food and the right partner. Among mammals, the elephant has the longest nose and the most olfactory receptors. Whether it also has the best sense of smell because of it, however, has not yet been researched.

Having a good sense of smell is vital for animals. The sense of smell helps them to orientate themselves, to detect dangers early, to find food and to choose the right partner. Most animals can therefore detect smells much better than humans can. It is not only mammals that have an excellent olfactory organ; reptiles, birds and fish can also perceive smells sensitively. The longest nose in the animal kingdom is certainly the trunk of the elephant. To everyone's surprise, even scientists, recent studies have shown that elephants also have the most functional genes for olfactory receptors, about 2000, while mice, rats and dogs only have half of that. In humans, only about 400 olfactory receptors are left. By the way, one of our favorite mammals has no olfactory receptors at all, that is, the dolphin. It replaces the sense of smell with an excellent ultrasound sensor system.

The number of different olfactory receptors says something about how many different fragrances the nose can recognize, but not how sensitive it is. Here, the sensitivity of the individual olfactory receptors and their fit (specificity) for certain fragrances play just as important a role as the number of olfactory cells. And here, there is a surprise winner: With about one billion olfactory cells, the European eel is currently the known leader.

Animal Super Noses

For comparison: The German Shepherd has 200 million, while humans still have just 20 million olfactory cells. Scientists have calculated that the eel's nose can detect a drop of perfume in three times the water volume of Lake Constance. This outstanding sense of smell helps the eel to hunt its prey in dark water, but, above all, also to find its way back to native waters for mating and egg-laying. And that, even though it travels thousands of kilometers on its migrations.

The eel thus impressively proves that smelling works not only in the air, but also underwater. On the contrary: It was in the darkness of the primordial sea that smelling first arose. There, where all life on this earth began.

That includes our life too, by the way. So, it's no wonder that we also, strictly speaking, can smell "under water", because, even in us, the fragrance molecules must first pass through a thick, watery mucus layer to the olfactory cells. One of the last, still living ancestors of all vertebrates is the hagfish. It has lived for 300 million years at depths of up to 2000 meters in almost complete darkness. Even this eel has 10 genes for olfactory receptors, which can still be found today in all fish, and even in us humans. The most advanced fish, such as zebrafish, have more than 100 different olfactory receptors at the very least.

As animals moved onto land, fragrances became increasingly important for their lives. They were carried by the wind over much further distances than the eye (and ear) could reach. Thus, they could warn of enemies and dangers, indicate food sources and water, and inform about possible mating partners. With the requirements, the number of olfactory receptors steadily increased. Only when the eyes learned to see and distinguish more colors and the eyes and brain became better and more complexly developed—as in primates and humans—did the sense of smell lose some of its importance. As a result, the number of olfactory receptors dropped from over 1000 in rats and mice, dogs and cats, to about 550 in monkeys and about 400 in humans.

Birds, it was previously assumed, rely on their good eyesight when hunting prey; after all, they do not have noses as we know them, but beaks. However, scientists from the Max Planck Institute have now observed storks deliberately flying to freshly mowed meadows so as to more easily locate frogs, snails, and other creatures. Simply spraying green leaf fragrances was enough to attract the storks. The researchers suspect that the sense of smell could also be important to other bird species in their search for food, because birds of prey such as buzzards and red kites also like to head for freshly mowed meadows.

In addition, recent research shows that fragrant landmarks play a central role in the orientation of migratory birds. However, it is not yet known how many olfactory receptors play a role in this.

Practice Makes Perfect

It is clear that the dog has the upper hand over humans when considering the number of olfactory receptors. However, olfactory performance can hardly be predicted based on these numbers. In odor identification tests, humans and apes often performed similarly well. Rather, it depends on how important scent molecules are for daily life. Humans find bananas great, as do monkeys; dogs don't care about them at all. No wonder that a dog, therefore, is less able to detect the smell of a banana than a human is. Banana peels, on the other hand, have a very special effect on other animal noses: Male mice can't stand their smell at all. They react by releasing stress hormones. The reason for this: Female mice who are either heavily pregnant or nursing make it clear to the males, with the same smell of acetic acid-n-pentyl ester, that they are not to be trifled with and intend to defend their young against any attack by a strange male. This because male mice are known to kill foreign offspring so that only their own genes are passed on within the group.

But back to the dog noses. Some of them are the purest super noses and allow the animals to provide valuable services to humans as guide dogs or in detecting diseases. More on this in the Chap. 11. However, not all dogs are equipped with a super sniffer nose. The short, flat, poorly ventilated noses of a boxer, with a small area of olfactory mucosa and correspondingly fewer olfactory cells, are by far less suitable for tracking as the long noses of bloodhounds or shepherd dogs. Their sniffing frequency can increase to up to 300 times per minute, and thus transport even the smallest amounts of scent into the nose for analysis. At the same time, once the air is inhaled, along with the scent molecules it contains, it collects in a kind of "olfactory niche" in the animal's nose. The scent molecules are thus able to remain in contact with the olfactory cells for a long time. In humans, on the other hand, the nose is emptied with each exhalation.

However, a major contribution to the super nose is made by the conscious engagement with scents. Many animals are born blind and are dependent on the recognition of scents from the first breath. They must find the mother's breast, recognize conspecifics, or orient themselves in the environment. While humans usually only perceive scents unconsciously and do not pay attention to them, it is completely different for dogs, cats, and mice. From a young age, they "see" their world with their noses, practicing, training, and educating themselves in sniffing. The examples of perfumers and sommeliers in humans shows that intensive and conscious engagement with scents and daily hours of practice can also provide us humans with super noses. We, too, have outstanding olfactory abilities, if we would only use them. Whether we could ever achieve the perfection of a sniffer dog, however, is questionable. Exact scientific data is lacking, because, so far, no human has followed tracks on the ground with their nose over long distances or systematically dealt with sniffing out explosives and drugs.

6

The Secret Messages of Pheromones

> Animals do not need words to communicate. The main thing is that the chemistry is right. With pheromones, they recognize the sex of their species, attract desirable sex partners, and warn of enemies. A special organ aids in communication.

In some respects, animals are to be envied. They sniff and stink happily to themselves and no one is bothered by it. On the contrary, smelling a lamppost or another animal's rear end is like exchanging business cards. When a male dog leaves a scent mark on a tree, he leaves valuable information: "I am young, strong, and potent, so get out of my territory! Females welcome!" Even the exact time can be recognized by other dogs based on the decay of the scents. However, the next dog may try to "overspray" this message with its own messages, so it can smell quite intense at such marking spots. Only fellow dogs understand this scent language. It exists alongside the learned and instilled scent evaluation. Neither cats nor other mammals perceive the messages, let alone humans. Only other dogs are initiated into the code, and only other dogs produce dog pheromone. In them, however, it triggers reproducible, always identical "compulsive reactions".

When animals warn members of their fellow species of an impending danger, everyone will immediately flee. If a bitch is in heat, the male dog requires neither good words nor treats from his master–he simply dashes off to do his natural duty. Rank order, fear, or territorial boundaries are also communicated. Animals have a special organ for perceiving such pheromones so as to understand the chemical messages. Named after its discoverer, it is called the Jacobson organ or–due to its location in the nasal septum (vomer)–the vomeronasal organ (VNO). While humans and great apes have deactivated this important additional scent organ during their development, it is still extremely functional in mice, cats, and horses. The VNO consists of a blind-ending, thin tube on both sides of the nasal septum, which is difficult for the air that the animal breathes to reach. And it works like a rubber pipette, sucking in air and, thus, pheromones. Some animals have developed a special breathing technique, the flehmen, so to speak, to press the "pipette" particularly hard. Horses, sheep, and especially camels pull their noses and upper lips high. Cats also suck as many scent molecules as possible into the VNO in this way. There are sensory cells specialized in pheromones, which are directly wired to deep brain regions (hypothalamus) via thin nerve threads. This also explains why pheromones always trigger the same, reproducible reaction in the animal. Pheromones are already important as early as the newborn phase. A special teat pheromone guides them to the mother's breast and triggers the sucking reflex. Again, cats have a different scent than rabbits, dogs, or mice.

No Sex Without Scent

Besides all other tasks, pheromones are primarily important as love scents. In the animal kingdom, there is no sex without the right scent. Nature uses it for species preservation and knows all sorts of tricks. For example, the boar has inexplicably ceased developing at a crucial point: His corkscrew-shaped penis impresses no sow. But this shortcoming is immediately remedied in a rather rustic way: With a scent that, in this case, rather resembles a knockout drug, because the boar uses the pheromones androstenone and androstenol for his sexual desires. He distributes them with his saliva to trigger so-called tolerance rigidity in the sow: She remains still until everything is over. However, the seductive scents only work at the time of her ovulation.

Humans find the pheromone androstenone disgusting, stinking, as it does, like urine, making boar meat practically unsellable. Small boars were therefore often brutally castrated before sexual maturity in the past; fortunately, castration without anesthesia is now prohibited. The scent of androstenone is also found in truffles—no wonder the sow seeks it and loves to eat it. It even occurs in men's armpit sweat, which does not make men more attractive to women's noses, but, as scientists have shown, it is significantly less stinky to women during ovulation, if at no other time. Animal sexual pheromones can also appeal to humans, like musk from the ox of the same name or civet from the civet cat, but they have no animal effects.

Animal scent communication even works underwater, and also ensures a successful love life there. Lobsters would never get closer without the love scent but would instead fight. Because female lobsters can only mate immediately after molting, when they are still without a hard shell, without scent protection, they would even be at risk of being eaten by male animals. The scent makes the male lobster gracious, and he will even allow the fair one to move into his safe cave for a few days.

In the world of insects, alarm pheromones play a major role. Through them, insects organize their social state, mark food sources and warn of danger. More on this in the chapter "The scent of flying sex machines". Even small aphids can alert their conspecifics during an attack. A small drop of the alarm pheromone will cause all aphids in the immediate vicinity to suddenly drop to the ground.

Conspecifics Are Warned

Plants, on the other hand, defend themselves against aphids and other pests in their own way. They, too, produce pheromones, in this case, fragrances, with which they both attract the enemies of the aphids and warn their neighboring plants. Tomatoes defend themselves through the production of solanine, which spoils the appetite of attacking caterpillars. If a tomato plant is attacked by spider mites, it uses a fragrance to attract predatory mites that, in turn, eat the spider mites. Conspecifics sense that danger is imminent and can start producing specific defense substances early on. Whether different plant species can communicate with each other is still largely unexplored. We do know from sage, however, that it emits whole clouds of fragrances that are also understood by tobacco plants. Even underground, plants communicate via a network of roots and fungal threads, the so-called mycorrhizal network. It primarily serves to improve nutrient supply.

Animals can sometimes be deceived by pheromone imitations from the plant world. Valerian and catnip, for example, can trigger states similar to intoxication, sexually arousing the animals, so that they roll in it and nibble on it. This explains why cats are so delighted with appropriately scented toys. The use of special feel-good and social pheromones can also help reduce stress and promote well-being. They originate, similar to the so-called DAP (dog appeasing pheromone) in dogs, from the female mammary gland or they manifest as fragrances from pheromone glands behind the cat's ears. They are actually produced during breastfeeding or "nuzzling" by the cat mother in order to calm her children. Therefore, nuzzling or cuddling by cats is not necessarily a sign of love, but rather a way for them to mark us, and thus take possession of us.

The opposite effect is caused by the alarm pheromones, which actually warn conspecifics, cause stress and induce flight. The so-called "Piss off" plant, an herb frequently used in the garden that is supposed to repel cats, seems to contain a similar scent mixture. These deterrent pheromones also emerged early in evolution. Even the goldfish produces them. They are particularly important in wild animals, with each species having an individual pheromone alarm scent. The scent of humans as enemies, on the other hand, is not the result of pheromones, but is a learned experience. It also triggers a flight response in many animals. Some hunters therefore know a very special deterrent: They walk around their shot game, leaving their own scent, and thus prevent other animals such as foxes or wild boars from approaching, even though they "smell the roast".

7

The Scent of Flying Sex Machines

> In insects, the nose is in the antennae. They detect scents and pheromones, which the insects use to communicate. Some butterflies have specialized in sexual pheromones. Their males are supposed to do just one thing: act as flying sex machines to ensure offspring.

With their antennae flies, bees, moths and ants can perceive all kinds of scents. They recognize food, dangers and also the scent of conspecifics. On the antennae, there are olfactory hairs, the sensilla, inside of which are olfactory cells with scent sensors, so-called receptors. Each olfactory cell is a specialist; it only produces one type of scent receptor. The wall of the olfactory hair is porous, so that the scent substances can reach the sensory cells from the outside. There, the chemical information is converted into electrical impulses, which are transmitted via the long nerve fibers of the sensory cells into the brain, which, in insects, is very similar to those in humans and vertebrates. The electrical pulses, in turn, generate odor-specific excitation patterns, by which the insects recognize the different scent substances.

In Bochum in 2005, we succeeded for the first time in activating an insect scent receptor with a scent substance, thus showing that the receptors—in contrast to those of vertebrates—work as a tandem, consisting of a scent-specific part and an electricity-producing amplification part. This increases the sensitivity enormously, so that a few scent molecules alone are sufficient to excite a cell.

In these investigations, we accidentally found, among the 60 olfactory receptors of the fruit fly Drosophila, the "marzipan" receptor, and were even able to block it with an anti-scent. It's a pity that we have not yet found it in humans. It could save many marzipan lovers extra pounds at Christmas.

The Language of Pheromones

On the antennae, there is also a second receptor family, the so-called pheromone receptors. They are extremely specialized, react only to a single scent substance and serve exclusively for scent communication among the animals themselves. The wiring pattern of the olfactory and pheromone cells is hardwired in the brain from birth, but in very different ways.

With the olfactory receptors, flies, bumblebees, butterflies and ants can locate food sources and also recognize plants, fungi or animal tissues that are suitable for egg laying. Some scents are repellent and lead to escape behavior because they signal danger or come from predators. Such behaviors can also be reinforced in insects through reward (food) or punishment.

In addition, insects, like mammals, have developed an elaborate system of chemical communication among themselves, the "language of pheromones". Each insect species speaks its own language. Social insects like ants or bees recognize the members of their state as friends and can distinguish them from enemy colonies. There are trail pheromones, marking scents, with which one can find the way back to the food source or lead others to it. And there are warning and alarm pheromones that individuals emit when they are threatened, to warn the other colony members or even make them aggressive. Thus, the sting of a single bee, wasp or hornet is often enough to soon cause hundreds of them to be on your neck. Because, when bees and wasps use their stinger for defense, they simultaneously emit a scent that is perceived throughout the hive: danger is imminent! The whole swarm will rush to the fight immediately. All ants are great specialists in terms of pheromone scent. They produce over 25 fragrances that they use individually or in combination for communication.

Specialists at Work

Finally, there are many pheromones used in brood care to signal that the offspring need food and, specifically, which food. Worker bees, drones, and the queen receive different foods. The most well-known is probably the "Queen Substance", her special food. This is a pheromone that inhibits the development of ovaries in the worker bees, and thus prevents the breeding of further queens. But it also serves as a sexual attractant during the nuptial flight when a bee colony swarms out. And, as everywhere in the animal kingdom, such sexual pheromones play a very special role.

In my scientific research at the Max Planck Institute in Seewiesen, I studied butterflies, more precisely: moths. The aim was to find out how the males can find their females in the dark. We discovered that, in the approximately 400 different moth (Noctuidae) species, the females each emit a mixture of specific sexual pheromones for the express purpose of attracting males of the same species. With increasing species numbers, this led to the complexity of the fragrance molecules becoming higher and the mixtures becoming more multi-faceted over the course of evolution. Accordingly, the receivers became more and more specialized. Thus, 90% of the olfactory cells of the male silk moth can only recognize sexual pheromones, but they can do so in the smallest concentrations and over distances of kilometers. Since these animals store large energy reserves from birth, they do not need food. Their only task is to find a female and reproduce. They are actually on the move as "Flying Sex-Machines". The fact that they also emit a specific sexual pheromone is solely for safety: the females are only supposed to allow copulation with males that emit the right "perfume". After copulation, many female insects suddenly lose their attractiveness and absolutely refuse to mate again. A pheromone scent is also responsible for this: a farewell greeting from the males, who spray their ladies with it.

8

Deception and Trickery with Irresistible Scents

> Some plants and animals lure with beauty, others seduce with irresistible scents. Certain substances even allow them to deceive rivals and drive enemies to flight. Humans use such scents for their own purposes.

For the survival of the species, one must make an effort. And if the natural attractions are not enough? Then one simply resorts to tools. Humans have perfected both small and large tricks. But animals and plants also know how to create optical illusions and false temptations: Their world thrives on attractants.

Some imitate the attractants of certain insects to lure them and achieve better pollination. or even to catch and eat them. The bumblebee orchid (Ophrys, Ragwurz) is a good example. With its flower, it not only imitates the hind body and color of a certain bumblebee species, but also mimics its sexual attractant. With this, it wants to improve its chances of pollination and optimal fertilization—within its own orchid species.

Digger bees, which are closely related to bumblebees, make it even simpler. They save themselves the production of scent by collecting the sexual perfume from the flowers instead, simultaneously using their visit to leave a fragrant footprint for everyone else: I was already here, there is nothing more to collect.

Scenting by the Clock

Not quite as specific, but no less effective, are plants that have precisely adapted the production of attractants to the flight times of the pollinating insects. Thus, the burning bush (Dictamnus) only emits its scent in the evening at dusk, when the hawk moths among the butterflies, which pollinate the flowers with their long proboscises, are active. Preferences for certain scents are also used to increase efficiency. This can be the honey-like scent of the linden blossoms, but also the scent of fermenting fruits of the cuckoo pint, which mainly attracts fruit flies. By the time the insect notices the deception, the plant has already attached its pollen, which is then carried by the fly to the next flower.

The fecal-like scent of the stinkhorn or the rotten meat stinking hawthorn, the foul smell of the carrion flower or the South American orchid (Satyrium pumilum) make the plants attractive to all insects that eat carrion (necrophages) or lay their eggs there. These can be flesh flies, but also houseflies and many beetle species. The scent imitation of the deceptive plants is so real and irresistible that the female flies believe it to be the perfect egg-laying place on animal carcasses. Taste is already a matter of dispute among insects.

But there are also insect-eating plants like the sundew, the Venus flytrap and the pitcher plant, which produce special fragrances and use them to specifically lure insects into the trap. Even more sophisticated are some male butterflies (like the cabbage white), which spray the female with a scent (methyl salicylate) after copulation, a clear signal for other males: Do not approach, oh unwanted, this female is already fertilized, so buzz off!

No Sex, Please!

This "anti-Viagra" for butterflies could, of course, also be used in the future to make all, even the still unfertilized females, unattractive in a region. Already in use in viticulture are sexual pheromones of such pests as the bark beetle and the grapevine moth. Pheromone traps, which are equipped with the corresponding sexual scents, are used to capture males in order to remove as many as possible from circulation. Additionally, the number of males captured per day serves as information for carrying out the use of insecticides in a targeted manner and limiting it to the shortest possible period. By now, we also know the sexual pheromones that female domestic pests use, such as those of food and clothes moths. Here, too, pheromone traps are used to attract and catch the males. However, one should always keep the window closed; otherwise, there is a risk that the sexual scent will also attract animals from outside into the apartment.

Due to climate change, we are also experiencing more and more mosquito plagues. Along with biting mosquitoes, disease-transmitting mosquito species from southern regions are now also native to us, carrying malaria, fever or sleeping sickness. The new scientific findings about smell also help in combating these animals. "Scent blockers" have already been developed to drastically reduce the animals' sense of smell. But it has also been researched as to why about 20 percent of people are strong mosquito magnets and are preferentially stung. It turns out that a genetic predisposition is to blame: Normal biting mosquitoes are particularly attracted to carbon dioxide (CO_2), and the more of it a person or animal releases through breathing, the more attractive they become to mosquitoes. The mosquito can smell its victim up to 50 meters away.

Particularly affected are pregnant women and people with blood type 0, at least, this is what is indicated by experiments with the Asian tiger mosquito. The influence of skin bacteria also seems to be crucial, as they produce different fragrances through their metabolism. The larger the population of microorganisms on the skin, the more susceptible the person becomes to mosquito bites. However, since human sweat is composed of more than 400 fragrances, it has not yet been possible to find the exact essence that repels or attracts mosquitoes. Some plants and herbs are said to have a deterrent effect on mosquitoes: lemon balm, basil and peppermint are supposed to drive them away, as well as sage, tomato plants and lavender. The latter also helps against moths, spiders and fleas in the closet. The downside is, depending on personal disposition, more or less effective anti-buzz remedies and the tormenting itch after summer evenings will probably accompany us for a while.

For the plants themselves, the most important significance of the essential oils lies in their protective effect against infestation by viruses, bacteria and fungi. They have optimized the composition of the essential oil mixtures for millions of years. Humans also benefit from this, and have been using this supermix for bacterial infections, fungal infestations and viral diseases since time immemorial, even in modern times, for Corona infections.

9

Nobody Smells as Good as You

> In our deodorized world, nobody wants to smell anymore. Yet body odors can be very attractive. They guide women to suitable love partners and signal to men when the conquest is worthwhile. All of this in the sense of nature: to preserve the species with a healthy gene pool. And possibly also for pleasure.

Every person smells different. This applies to both their body odor and to the ability of their nose. It quickly knows which scents we find repulsive or attractive, whether we can smell someone or not. Whether snub nose or eagle model—shape and size are completely irrelevant. Every person evaluates scents individually, depending solely on the memories of the nose bearer and their genetic equipment. Actually, we should enjoy our body odor and trust nature. It reliably takes care of its favorite concern: to equip us perfectly for survival and to ensure the most successful reproduction of the species. To this end, it tries everything, so that genes that fit together optimally also come together and that they meet at the right time. Long before we humans developed high-tech devices for gene analytics, nature equipped our noses with it: We can "smell" the genes of fellow human beings through their body odor.

But what do humans do? They utterly overreact to physical scent messages. Every hair that could serve as a scent distributor is removed or shampooed, every scent molecule on the body mercilessly showered off. As if that weren't enough, they then interfere with nature through the use of foreign scenting. Men, for example, would apparently prefer to smell like musk, in other words, like the groin area of the deer of the same name, or like the anal area of the civet cat, rather than themselves.

Tempting Messages

Encased in sweat are four components that we can perceive with the nose: the smell of sweat itself, the smell of food (meat-eaters smell different than vegetarians), the individual smell and the messages of the pheromones, scent messages that every other human understands and reacts to immediately.

What the modern successful person fears is the typical rancid-fatty sweat smell. In fact, all humans have three to four million sweat glands, which can produce up to ten liters of sweat every day, individual mixtures of salt, ammonia, fats, sugar, acids, and fragrances. Humans sweat, and the sweat provides cooling through evaporation. Here is where the misunderstanding comes in: Fresh sweat doesn't stink at all. The rancid cocktail comes not from us, but is only produced by our companions, in other words, the bacteria and other microorganisms on our skin. They break down the long-chain fatty acids that originate in the sebaceous glands into shorter chains, producing the terribly stinking butyric acid and the biting formic acid. You can also sweat as a result of excitement or because you're scared. Then, sweat formation is controlled by hormones that particularly influence the sebaceous glands. Over the course of our lives, our body odor changes. Babies still smell enticingly sweet, while, later, the scent notes change, after which some are then well advised to let their jogging shoes air out on the balcony.

The negative evaluation of sweat is shaped by our upbringing. "Child, wash yourself, you stink," our mother instructs us. In the past, people were often far more relaxed concerning body odors. And—it should not be concealed—some scents can also be attractive. Cleanliness was even avoided at the end of the eighteenth century because people feared losing their own seductive power. Legendary in this respect is Napoleon's request: after months of professional absence from home, he wrote to his Josephine days before his arrival: "Do not wash, I am coming". Biologically, this is quite sensible, because sweat contains many chemical messages for our fellow human beings.

The Decisive Body Odor

The smell of sweat is, as mentioned, only one component of our body odor. The other is the inherent smell, which is largely determined by our immune system. Every person is therefore their own perfumer: they produce an individual perfume that is determined by their genes. This individual smell makes people unmistakable; it is, so to speak, an "olfactory fingerprint". Dogs recognize us immediately by it. And the Stasi, during the days of the GDR, also took advantage of the fact that every person has a very individual smell. They collected sweat samples from detainees. Either during house searches or after sports, socks, shoes and laundry were confiscated. Another method was impressively demonstrated in the film "The Lives of Others", a dramatization of life in the GDR: After hours of sweaty interrogations of prisoners, the covers of their seat cushions were put into preserving jars to vacuum pack them for preservation, so that they would be presented to sniffer dogs if necessary. This way, even after a suspect was released, the authorities always had so-called "comparison materials" available and could quickly identify him based on scent traces. Even today, hundreds of scent samples are on display in the German Spy Museum in Berlin.

Why does nature go to such lengths with individual smells? Of course, it is solely for the purpose of better species preservation! Unconsciously, women react to the information of the individual smell when they are looking for a father for their children. He should bring genes that differ as much as possible from their own. Experiments with men's T-shirts have shown: The more the man's body odor differs from their own, the more attractive he appears to them. In this way, women ensure a well-mixed gene pool that equips the offspring with a more stable immune system, and thus better health. No wonder that, in the scent-free world of online dating sites, genotyping is already being offered—not very romantic, but similarly effective. Also interesting: During pregnancy, a woman's scent preference changes. For the upbringing of the children, she relies more on men with a similar body odor, i.e., her own family. The sensitivity to scents also changes, especially in the first weeks of pregnancy. Often, women therefore cannot stand certain smells at all, and may become nauseous from the smell of coffee, gasoline or certain foods.

The Best Perfume for Women

Men don't care about the gene pool. They are less concerned with the quality than with the quantity of offspring. They are hardly interested in the genetic makeup the future mother brings. The main thing is that their own genes are passed on, and they seek as many opportunities as possible to do so.

The scent of a perfume that the man associates with positive memories, however, can increase the attractiveness of women. So, is there a scent that makes women attractive? Swiss researchers wondered about this, and subsequently found out: Yes, some women smell better to men than others. And no, it's not because of the shampoo or the perfume. Their hormonal status is crucial. The composition of the sex hormones changes over the course of the cycle. Women smell more consistently interesting to men when they produce more estrogen on their fertile days. The more estrogen, the more attractive—that's how the result can be summarized. This means: Men can indeed receive information about a woman's potential fertility through their noses and will invest their energies and juices accordingly.

By the way, strippers have long been aware of the increased sex appeal during their fertile days. American scientists studied the influence of the female cycle on the earnings of the dancers and, indeed, discovered that they received twice as many tips on their fertile days as usual. Finally, a scientific study that thrilled all participants.

10

Cold Sweat and Baby Scent

> Humans are rather reserved when it comes to the language of pheromones. However, we too can send and receive scent signals that are understood by other people and trigger certain, always identical reactions, completely unconsciously and without us being able to control or regulate them. A somewhat unsettling thought.

As far as the messages of pheromone communication are concerned, one must say: Humans are only sub-optimally equipped for this. But: We still possess about 10 receptors that guide us safely through the wild mix of the everyday molecular cocktail. This informs our noses about fear, stress, aggression or cycle status; it triggers trust or compassion. The newborn recognizes the mother and finds the source of milk, namely, the mother's breast, blindly.

Like animals, we apparently produce warning signals when we feel fear and pleasant scent messages when we are happy. This was discovered by an American psychologist, who showed moviegoers both comedies and horror films. The visitors then provided sweat samples, which other test subjects clearly identified as "joy" or "fear". Similar experiments with dogs confirmed this. For a dog's nose, it is no problem at all to smell whether the human was watching a sex, love or war film.

Human fear sweat is unmistakable and unconsciously triggers the same reactions in all people: one becomes more attentive, active, but also somewhat anxious and empathetic. This was demonstrated by psychologists at the University of Düsseldorf. Unfortunately, neither the scent nor the receptor for it is known. When parents lovingly care for their offspring, it is because babies encourage their parents to do so with their scent. Mothers not only easily come to know their babies by this scent but can even sniff out their own offspring within a group of children. This works until the age of nine, when early puberty begins to change body odor. In the meantime, the magic scent of babies has also been identified in the lab. Its name is hexadecanal. And it has another advantage: it simultaneously reduces human aggression.

The smell of newborns can stimulate the brain as effectively as anti-anxiety and depression medications. This was discovered by Swedish scientists who are now intensively working on a nasal spray with baby scent that could be used as an antidepressant.

The Smell of Trust and Friendship

Conversely, mothers apparently send out messages that signal to the baby: You can trust me, you are safe and well cared for with me. And women unconsciously give men hints about the time of their ovulation. They simply smell more attractive, so that men—also unconsciously—begin their conquest.

A woman's tears can also control male behavior. Neurobiologist Noam Sobel describes the effect in a new study: Female tears, according to Sobel, lead to a drop in testosterone levels in men, and thus to lower sexual arousal.

The fact that we rely on our noses and our genes when choosing a partner has been known for some time. The Israeli Weizmann Institute of Science headed by Sobel has provided completely new insights: Friendships are often based on body odor. Friends must be reliable, often exhibiting personalities that are similar to our own, and may even be genetically similar to us. The researchers examined the body odor of friends and, indeed, determined: Good friends smell similar to each other.

It is our habit to try to figure out how to assess our counterpart, often unconsciously. After every handshake, the researchers of the same institute discovered, people tend to bring their hands close to their noses and smell them. In this way, every day, we collect data from the body odors of our fellow human beings. Similar to dogs, these scents inform us about the health status and emotional state of these people. This is probably also the reason why people who have lost their sense of smell feel greatly impaired in their social relationships.

At Ruhr University in Bochum, we were able to decipher the activating scent for one of the human pheromone receptors: Hedion. It is a fragrance that occurs in jasmine and that must have a chemical equivalent in humans. MRI studies showed that the same small region in the hypothalamus always reacts with increased brain activity, up to ten times more in women than in men. Together with a behavioral economist from the University of Cologne, we then examined the behavior of people and discovered: When Hedion was in the room, they reacted with significantly more trust during games involving rewards and with more distrust in games involving punishment.

Another type of pheromone receptor recognizes amines, fragrances that are mostly formed by decaying bacteria and that smell like dead fish. Mice use them to recognize sick animals. Female mice with this scent are avoided by males. Humans also still have some receptors of this type, as we were able to show. Whether the rotten-fishy amine stench in inflammations in the oral and vaginal area reduces reproductive activity, we do not yet know, but it certainly doesn't promote physical contact. We have already found and patented a blocker against this smell: to reduce "bad odors".

11

Diagnostics with the Nose

> Some diseases reveal themselves through characteristic smells. They make the breath sweet, the sweat sour, or the urine sharp. Cancer cells, malaria and epilepsy also change body odor. So far, only trained animal noses and special tests can detect these early warning signals of the body.

Modern laboratory medicine knows everything. It knows blood and hormone levels, can identify germs and fungi, and can check all organs. But sometimes, the use of a fine nose is enough to recognize the first symptoms of disease. Sour or sweet? Rotten or fresh? Musk, apples, or ammonia? With certain diseases, the body produces metabolic products that are excreted through sweat, breath, or urine, and have a strong inherent smell. Some of these smells could be helpful as early warning signals and might even make complicated tests unnecessary in the future.

It's easy to guess that bad breath indicates caries or periodontitis. In up to 90 percent of cases, the cause of bad breath is in the mouth area. It's not so obvious that a sweet breath suggests tonsillitis. The bacteria that cause the inflammation also cause the formation of pus pustules. These, in turn, emit the sweet scent. If it also smells rotten, it could suggest pneumonia or—as was often the case in the past—diphtheria. Fortunately, the life-threatening diphtheria has largely been eradicated in our country thanks to the high vaccination rate in toddlers. Acidic breath—caused by an inflamed stomach lining that produces too much stomach acid—is encountered more frequently. The cause can be bacteria, tumors, or even stress.

The two diseases that most often cause bad breath are diabetes and kidney disease. A fruity smell (apple), sometimes also the slight smell of acetone-containing nail polish remover, is typical for diabetics. The urine can also smell like malic acid or acetone. The metabolism is disturbed: The body does not have enough insulin and is flooded with fatty acids, leading to the formation of malic acid.

When Sweat Smells Like Urine

Sweat and breath that smell like urine or ammonia can indicate kidney weakness or even acute kidney failure. If the kidney is not working properly, the body does not excrete toxins through the urine. This causes urea to enter the bloodstream in larger quantities, and it is subsequently sweated out through the skin and exhaled through the lungs. A biting ammonia smell can also indicate a sick liver.

In contrast, a bladder infection, which makes the urine smell strong and sharp, seems rather harmless. But many women know how painful such a bacterial infection can be. If the urine smells sweet, a hereditary disease may be present. It can be detected in newborns and is named after its smell: maple syrup disease.

Those suffering from hypothyroidism smell like vinegar. The reason for this is the slowed metabolism. Acids are formed in the body and are often sweated out at night. One of the most common congenital hereditary diseases is phenylketonuria. In this case, the amino acid phenylalanine accumulates in the body and disrupts mental development. The sweat and urine of an affected child emit a musky smell. But the situation doesn't always have to be so dramatic: even a common cold with fever can significantly change body odor.

Typical smells are also known in forensic medicine. The typical bitter almond smell of cyanide poisoning can still be detected in the deceased, as can arsenic poisoning, which smells like garlic. If the deceased suffered from typhus, the examiner will detect the smell of freshly baked bread.

Many diseases would be more curable if they were detected early. Do cancer diseases possibly also produce smells that could be detected by sensitive noses? This question was first pursued by a research team in California. They trained dogs to recognize cancer patients based on their breath samples. In a short time, the dogs achieved hit rates that were almost as good as those of elaborate laboratory tests. It is now considered certain that dogs can detect lung, breast and bladder cancer based on breath and urine samples. The dog trainer of the Israeli company "Dogprognose" promises that his dog Timi is correct in 95 percent of cases. Attempts are also being made at Max Planck Institutes to imitate dog noses and develop biosensors that can detect tumor diseases. They could be used, for example, in a doctor's waiting room so that the doctor may know early on if a patient with cancer has come for an appointment.

Dogs and Rats as Lifesavers

Dogs apparently also sense the olfactory changes of an impending seizure in epileptics. Patients have long reported their companion dogs warning them of a seizure by behaving restlessly. In a scientific study in France, this ability was tested with three female and two male dogs of different breeds. They came from a training center for companion dogs and could already detect diabetes and some types of tumors. Odor samples were taken from the patients with cotton balls, and the individuals were also made to breathe into a bag. During a learning phase, the dogs were repeatedly presented with samples from epilepsy patients and healthy people. It was confirmed: All the dogs involved were able to identify epilepsy patients who they did not previously know. The researchers were surprised by the accuracy of the results, but also admitted that the number of test dogs was very small. It has also been possible to train dogs to recognize Covid patients. The dogs were able to identify infected people based on saliva and respiratory secretions. With 80 percent certainty, they were almost as reliable as a conventional rapid test. They were also able to do this in later tests with sweat and urine samples. The latest studies led by the University of Veterinary Medicine Hannover now show that dogs can even recognize Long-Covid patients. The samples came from patients at the Hannover Medical School, where the virus was no longer detectable by PCR test. Here, their hit rate was as high as 90 percent.

The most common example of this phenomenon is diabetic alert dogs. They recognize the first signs of a diabetic coma in severely diabetic patients, so that the patient still has time to take their insulin. Diabetic alert dogs are trained assistance dogs that undergo training for 18–24 months. They can save lives daily, preventing coma and seizures. High blood sugar levels are indicated by good diabetic alert dogs from 170 mg/100 ml. In this way, the diabetic can take carbohydrates or inject insulin in time to counteract hypoglycemia or hyperglycemia, as well as reducing the risk of secondary diseases. The diabetic alert dog warns its diabetic by nudging them or placing its paw on them, for example.

Scientists at the TU Braunschweig have found that rats, with their fine noses, can also be helpful in detecting diseases. The African giant pouched rat is now being trained to recognize tuberculosis. In Africa, only about half of TB cases are detected, as the diagnosis is expensive. The so-called "HeroRats", which are used by the non-profit organization APOPO, have already achieved hit rates of 75 percent, and could thus be an alternative to conventional tests.

Super Nose Discovers Parkinson's

So far, there is no test for early detection of Alzheimer's and Parkinson's based on scents. It is only known that the ability to smell decreases years before the onset of the disease. Often, sudden smells appear, for example, of fresh bread, that do not fit at all. A neurologist from Oregon reports that he himself experienced such phantosmias for years, after which they disappeared. He was 55 years old at the time and feared that he might be developing Parkinson's. A DNA analysis then revealed that he carried a roughly twelve-fold increased risk of Alzheimer's.

There were no indications that these diseases themselves produce certain smells and no animal nose had detected any. That is, until an Englishwoman with a super nose made headlines: Joy Milne can apparently smell Parkinson's even before the first symptoms appear. She reports that her husband Les had smelled of musk for ten years before the disease was diagnosed in him. At a meeting with other Parkinson's patients, she later discovered: All these people smelled the same. Joy turned to a doctor at the University of Edinburgh and reported her suspicion. He conducted a test: Joy was asked to sniff twelve worn T-shirts, six from Parkinson's patients, six from healthy individuals. Indeed, Joy was able to identify all of the T-shirts belonging to the sick individuals, while also picking out one from one of the volunteers. Shortly thereafter, Parkinson's was also diagnosed in this previously healthy individual. Scientists from the Universities of Manchester and Edinburgh continue to work with Joy Milne as of this writing. First research result: The distinctive smell seems to be associated with sebum, a skin secretion. Sebum is—like other molecular compounds—produced in greater quantities in people with Parkinson's. Now, the researchers are looking for the individual chemical compounds that produce the smell. With Joy Milne, they continue to work on the diagnosis of Alzheimer's and cancer and on a diagnostic test for the possible detection of tuberculosis.

12

Smelling with Skin and Hair

> That which long sounded unbelievable is increasingly occupying scientists all over the world: Olfactory receptors have spread from the nose throughout the entire body. They have now been detected in all tissues: in the skin, in the hair roots, in the intestine, in the heart, and even in sperm. By the way, they are particularly fond of the scent of "lily of the valley".

The nose remains our favorite and only olfactory organ. Olfactory receptors do exist throughout the body, but they have nothing to do with "smelling", rather performing many other important tasks. They react to fragrances that enter the blood in high concentrations through breathing, contact with the skin, or ingestion, and are distributed from there throughout the body to the brain. Added to this are the many fragrances produced by microorganisms on and in our body, such as on the skin or in the intestine. Sweat and "fart" smells give us an impression of this.

Fifteen years ago, we were able to prove, for the first time in our laboratory at Ruhr University Bochum, that olfactory receptors also occur outside the nose and are actually still functional. This was no easy task, as their discoverers—two Nobel laureates—had previously claimed the opposite. Linda Buck and Richard Axel, who received the Nobel Prize in Medicine in 2004, had assumed that olfactory receptors only occurred in the nose. We were successful when we studied the miracle of human reproduction and asked ourselves: How do millions of tiny sperm orient themselves in the complete darkness of the female body? And find their target so reliably? In the process, we discovered that 15 different olfactory receptors that we knew from the nose were also in the sperm. The first of these reacted to a synthetic lily of the valley scent. If it smelled like "lily of the valley", the sperm knew only one direction:

towards the source of the scent, and quickly at that. The fact that we were then able to detect the corresponding fragrances in vaginal secretions was a rather exciting observation that inspired many scientists worldwide to further research. It was even more exciting when it became possible to find a blocking scent for individual olfactory receptors, and thus, so to speak, "hold the nose" of the sperm.

Contraception with anti-scent—what a possibility! Just as fragrances fit like a key into the lock of the olfactory receptor to activate it, the anti-scents manage to close the lock precisely with the same key principle. Scientists call such blocking substances antagonists. Would they also work on the olfactory receptors in the nose? We expected this, and were actually able to experience it impressively. In an experiment, we asked a number of test subjects to enter a room full of lily of the valley scent, or to be precise, its chemical variant. And lo and behold: when we blew the scent blocker into the room, the lily of the valley scent suddenly disappeared. None of the subjects could perceive it anymore. Could scent blockers also make other smells disappear? We now assume that each olfactory receptor has its own blocking scent—a wide field of research that could make everyday life easier for us in many situations.

Hair Roots Recognize Sandalwood Scent

We initially focused on the search for olfactory receptors in the human body and examined the skin, because our skin cells naturally often come into contact with fragrances: when peeling an orange, applying a cream, or through scents in sweat and those produced by the microbiome. We immediately discovered over 20 different olfactory receptors. "Smell turns up in unexpected places", the New York Times wrote in their report on our research results.

One of the olfactory receptors that we have now examined more closely reacts to Sandalore, a synthetic sandalwood scent: Skin cells (keratinocytes) multiply and move faster when they come into contact with sandalwood scent. This does not apply to natural sandalwood, but only to the synthetic scent Sandalore. Wounds heal 40 percent faster and the skin regenerates better—even in old age. These findings are now also being used therapeutically, e.g., when larger wound areas need to be treated after burns.

We were also able to find this olfactory receptor in hair root cells. A clinical study with Sandalore lotion, which was applied to the hair roots for 12 weeks, showed that this scent extends the lifespan of the hair by about 20 percent. Finally, fuller hair. There are now also preparations for skin and hair that exploit these findings therapeutically. We know little about the other olfactory receptors in the skin, especially as to whether they might even have negative effects. Until this is researched, one should not spray perfumes on the skin, but rather on hair and clothing.

It would seem that the olfactory receptor named OR2W3, which occurs in at least 20 different human tissues, including the brain, lungs, thyroid, and white blood cells, is particularly widespread in the human body. Unfortunately, no activating scent for this olfactory receptor is known yet. In various organs, olfactory receptors enhance or prevent cell growth, as well as influencing communication between cells and, apparently, also the production of hormones such as serotonin. In the pancreas, the serotonin production also influences insulin formation. In the intestine, serotonin acts as a regulator for intestinal peristalsis and the secretion formation, which can result in diarrhea or constipation.

These examples show the importance of olfactory receptors outside the nose, including in regard to physiology. But they are only the "tip of the iceberg". Unfortunately, we only know the activating scent for 20 percent of human olfactory receptors. Although we know exactly, through new genetic technological advances, which olfactory receptors occur in which body cells, as long as we cannot activate the olfactory receptor, we cannot study its effect. There are tissues in the body, e.g., at the entrance of the cervix or in the ureter, where olfactory receptors even occur in much larger quantities than in the olfactory cells of the nose. Why and for what purpose? The next few years will bring many exciting new insights.

13

Scents as Therapy Aids

> Herbs help with stomach aches and coughs. Now, we know the reason: In the gut and other organs, the same olfactory receptors work as in the nose. Some of them react to fats, and are located in the kidney and the heart. This represents a possible new approach to regulating blood pressure.

Sometimes, fragrances do not work through smelling, but one must rather inhale them, eat them, or rub them into the skin. Many people suffer from high blood pressure. Not a particularly romantic idea: Our heart's blood is not only shaped by passion, but also by a lot of blood fats. In the gut, enzymes and microorganisms help to break down the fats from food into different lengths of fatty acids and distribute them as energy suppliers via the blood throughout the body. For fatty acids, there are different olfactory receptors in the nose, with which we can distinguish olive, coconut or sunflower oil by smell. We have found the same "fat receptors" in the kidney, and now also in the heart. In the kidney, they welcome short-chain fatty acids and are—although not always to our advantage—involved in the release of the blood pressure-increasing hormone renin. Our food intake can thus influence blood pressure—with the help of olfactory receptors. If you destroy these olfactory receptors in mice, it leads to permanently low blood pressure, as shown in experiments at Johns Hopkins University. Could we therefore fight high blood pressure by blocking the fatty acid receptor in the kidney? After all, from the nose, we know: There is always a specific activating scent, but also a blocking one.

We were able to detect the olfactory receptors of the heart in human "mini-hearts" grown from skin cells or embryonic stem cells, but also in heart tissue from surgery, and show that certain fatty acids, in this case, the medium-chain ones, activate them, and thereby slow down the heartbeat (frequency) and reduce the heart's power. They therefore have negative effects. Especially in diabetics, we found this type of fatty acid significantly increased. Here, the use of an anti-scent against this olfactory receptor could be helpful, as we could demonstrate experimentally.

Such an anti-scent could also possibly help in the fight against arteriosclerosis. This was recently discovered by American researchers. They were able to show that, in cells of our immune system (macrophages), an olfactory receptor exists that is activated by octanal, a citrus scent. The activation of this receptor causes the release of inflammation-promoting messengers, which, in turn, accelerate arteriosclerosis. There is also a blocking scent for this, namely, citral, which smells like lemongrass. This scent blocker could therefore actually be a novel therapeutic approach in the treatment of arteriosclerosis.

Herbal Scents for Digestion

We also constantly take in scents through breathing and eating that are distributed throughout the body with the blood. Plant experts, as well as herbal liqueur lovers, have long known that spices stimulate the stomach and intestines, but can also calm them. We were therefore not surprised to find olfactory receptors for various spice scents from caraway (carvone) or cloves (eugenol) in the intestinal cells. Their activation triggers the release of messengers there, which accelerate or slow down intestinal peristalsis, and thus our digestion can be controlled. A digestif after eating—recommended for "medical reasons", even if not by prescription. The gut flora even produce many of their own fragrances, as can be easily noticed in "winds" and excrements, whose scent varies depending on the food.

Fragrances Against Respiratory Diseases

With every breath, fragrances come into contact with the bronchi and lung tissue; additionally, the moist, warm environment is an El Dorado for various scent-producing bacteria. No wonder that we have found many different olfactory receptors there as well.

We were able to examine two of them more closely. They are located in the smooth muscle cells that encircle the bronchi like a ring. The receptor OR2AG1 was particularly interesting, as its activation by the fragrance compound amyl butyrate (with a smell like pear and apricot) leads to a relaxation of the bronchial muscles (and thus an expansion of the bronchi), allowing more air to enter the lungs again. In diseases such as asthma, allergies, and COPD (chronic obstructive pulmonary disease), in which the bronchial muscles constrict and the airway is reduced, inhalation of the fragrance could allow more air to enter the lungs, similar to the administration of cortisone. In allergies, the release of histamine due to inflammation often leads to additional muscle contraction, but even this can be completely overridden, researchers would say "overruled," by the fragrance compound.

The second scent receptor (OR1D2), which we found in the smooth muscles of the bronchi, is activated by cyclamal (with a smell like lily of the valley) and causes exactly the opposite: a contraction of the muscles, and thus a narrowing of the bronchi. It exacerbates shortness of breath in patients with obstructive lung diseases, allergies, or asthma. This shows how important it could be to know all olfactory receptors and their functions so as to utilize the positive effects and avoid the negative ones.

This also concerns our latest research results in cooperation with the Bergmannsheil Clinic in Bochum, where we examined immune cells, so-called macrophages, in the airways. These defense and scavenger cells play a role in lung patients with asthma, allergies, or COPD. In the lung mucus, they ensure the removal of bacteria, viruses, and environmental pollutants from the inhaled air. At the same time, however, they can also contribute to inflammatory processes in lung diseases such as COPD. When we examined these lung macrophages, we found a multitude of receptors.

We selected two: the receptor OR2AT4 (Brahmanol), already found in epithelial cells, and OR1A2 (Citronellal). We already knew the activating scent of both. When we added the two fragrance compounds to the isolated human macrophages obtained through bronchoscopy, we observed two things: a strong decrease in the release of various inflammation-inducing substances and a reduction in the "scavenging" properties of the macrophages. Here, too, inhalation of the scents promises high therapeutic potential.

In extensive studies, we also examined lung epithelial cells from patients with chronic inflammatory obstructive airway diseases such as COPD, which were steroid (cortisone) resistant, in cooperation with the Bergmannsheil Clinic in Bochum. There, too, we discovered some olfactory receptors. We were able to decipher the activating scent of two: OR2AT4 (Brahmanol) and ORJ2/3 (Cinnamaldehyde). When the lung cells were stimulated with cinnamaldehyde, the release of inflammatory mediators, substances that initiate or maintain an inflammatory response in the body, was reduced. The cell division rate increased, and wound healing was improved. Brahmanol had a significantly lesser effect but did significantly increase wound healing. Especially when cortisone proves unhelpful, the use of cinnamaldehyde could have important therapeutic effects.

Overall, a new pharmacological toolkit is thus available that could be very valuable, especially in those diseases that do not respond to cortisone. Since the affected cells can be optimally reached by inhaling the fragrance compounds, and thus the receptors can be stimulated, application would also be easy and effectively feasible.

14

Fragrances Against Tumor Cells

> Scent receptors exist throughout the body. And they also play a role in diseases, even in cancer. Because tumor cells can also "smell". Some grow more slowly, others even die off, an exciting discovery for future therapies.

In recent years, we at the Ruhr University in Bochum and other laboratories worldwide have collected many scientific data that prove: In all tumors studied so far, the number and pattern of olfactory receptors compared to healthy tissue is significantly changed. A lot of such olfactory receptors were found in the tissue of prostate cancer tumors, in liver, lung, bladder, and colon cancer, as well as in leukemia cells. If these receptors are activated by a certain fragrance, it can have many biological effects on the cell. The cells can be induced to divide less, move or even die off earlier and go into programmed cell death. Cancer cells that die faster—a glimmer of hope for patients? At the very least, it represents a new innovative therapy approach.

For an olfactory receptor that occurs in large quantities in colon cancer cells, we were able to determine that it is activated by a fragrance from the privet flower (troene). To investigate its effect precisely, we brought tissue samples from tumor patients into contact with it. The result was clear: The cancer cells died off or grew more slowly. Also, the migration of the cells was greatly reduced, which makes the formation of metastases more difficult.

Something similar happened with bladder cancer: In cell culture studies with cancer tissue from patients, it became clear that an olfactory receptor occurs much more frequently that reacts to fragrance components from the natural sandalwood aroma. So, it is not the synthetic sandalore that is needed in this case, but the real sandalwood scent. When coming into contact with these fragrances, the cancer cells divided less frequently and were not as mobile. Interestingly, there were already reports over 100 years ago that sandalwood was used against bladder cancer. In the case of liver cancer, it is the citrus scent that has such an effect. And with leukemia cells, it is a fruit scent. Now, clinical studies are still needed so that patients can benefit from it.

New Approaches for Diagnosis and Treatment

Olfactory receptors can also be helpful for the diagnosis of diseases. A few years ago, the olfactory receptor for "violet scent" was discovered in prostate cancer in such quantities that it can now be used as a tumor marker to distinguish healthy prostate cells from cancer cells. There is also a tissue-specific olfactory receptor in women's breast cancer.

Since it only occurs outside the nose in mammary carcinoma cells, one can recognize these cancer cells anywhere in the body. Unfortunately, the activating scent is not yet known, so its possible significance for therapeutic use cannot be investigated. New studies even found olfactory receptors in the blood and urine, thus opening up completely new, gentle options for early diagnosis. This procedure is called a "liquid biopsy". We were also able to detect the sandalwood receptor in large quantities in clinical studies in the urine of bladder cancer patients, and thus diagnose the disease early. There are now initial experiments with biosensors in toilets that respond to olfactory receptors in tumor cells and warn the user before they feel symptoms.

New research results from the examination of brain tissue in patients with neurodegenerative diseases such as Alzheimer's and Parkinson's have found large differences in the occurrence of olfactory receptors compared to healthy brain tissue. An exciting discovery in the quest to detect such diseases, but perhaps also an opportunity to open up new approaches for therapy. It has long been known that, in these diseases, a decrease in the sense of smell occurs almost 10 years before the first symptom appears, and therefore smell tests can provide a helpful early diagnosis.

The potential of the extranasal olfactory receptors has thus far not been sufficiently explored. And while we buy pharmaceuticals, food additives and cosmetics for healing or improving body functions today, in 20 years, it may well be olfactory wonder pills that end up on our shopping list, as fragrance imitators or olfactory receptor blockers, finding therapeutic use and providing health and increased well-being.

15

When Scents Go to Our Heads

> Through the lungs, the skin, or the stomach, scent molecules travel into the blood, and thus directly to the brain. There, they are received by special receptors and their effects unfold: They make us tired or alert and protect against motion sickness. All without conventional medication.

Scents that we perceive with our noses are evaluated very subjectively. We associate them with pleasant memories or unpleasant experiences. And the more intense the positive or negative emotion, the more stable the storage. The things that we experience usually have a smell, even if we don't consciously remember it. When it reappears, we also remember the experience. This is how scent preferences are formed unnoticed. The scent information and the emotions experienced with it are stored firmly as memories in the hippocampus, i.e., in the brain. They reach—unlike visual impressions—directly into this center for feeling and memory via a single switching point. Thus, a scent can trigger different effects in each person, depending on the situation in which we first encountered it. We have known about these paths of scent perception for some time.

But now, scientists have shown that scents can also continue to have effects in people who have completely lost their sense of smell due to illness or accident, and then even reproduce the same effects each time. How is this possible? Especially without functional olfactory cells in the nose? Researchers have shown that scent molecules that we inhale, eat, drink, or rub on the skin can also directly enter our blood, and thus be transported throughout the body. In this way, they come into contact with all of the cells of our body tissue—from the periphery to the brain.

In the outer membrane of all cells, but especially nerve cells, there are various receptor proteins, e.g., sensors for temperature, pH value, pressure, electrical potential, or hormones and neurotransmitters. Especially in regard to neurotransmitters, it is known that their function can be greatly altered by various chemical substances (drugs)—either becoming more or less sensitive.

GABA Receptors for Restful Sleep

Our work shows that scent substances that are distributed in our body via the blood also play an important role as modulators of neurotransmitter receptors. In this way, they influence physiological functions, as well as our behavior. In contrast to the subjective effects of scents through the nose and their activation of brain areas, the effects of scent substances in the blood, purely pharmacologically conditioned, are reproducible and are the same in every person.

In recent years, we have mainly dealt with the so-called GABA receptor in the human brain. Its name comes from the neurotransmitter gamma-aminobutyric acid, which activates it. It is produced by brain cells and is always released when the activity of neighboring brain cells is to be inhibited, for example, when at rest, in a relaxed state, or during sleep. The GABA receptor can be modulated by various chemical substances. They cause the natural neurotransmitter to have a stronger effect, making the person more tired or relaxed. This includes all classic sleep, calming, and anesthetic drugs, such as Valium, barbiturates, and propofol, but also alcohol, all of which reach the brain via the blood and have a sleep-promoting, calming, anxiety-relieving, or narcotic effect there. We have now identified more than 30 fragrances that also dock onto the GABA receptors, and thus act as calming or sleep aids. Fragrance components from the gardenia flower (vertacetal) and lavender (linalool) are among them. Lavender oils can be taken as capsules, inhaled, or rubbed on the skin to reach the blood vessels, to produce a sleep-promoting effect in the brain. In some cases, the fragrance is even more potent than Valium, the most well-known sleep aid, which also works through the activation of GABA receptors.

However, if the GABA receptors are blocked, for example, by menthol from mint, cineol from eucalyptus or beta-asarone from the calamus plant, the person remains awake and active. This was already known to the ancient Egyptians. They used calamus in the incense of the temples for invigoration and were also familiar with the calming scents from frankincense and various flowering plants.

Tonic Water Against Leg Cramps

An interesting fact for cruisers and other travelers who unexpectedly run into rough seas, or for people who get sick while driving, is the effect of fragrances that influence the neurotransmitter receptor for serotonin in the brain. This is mainly responsible for travel sickness and seasickness. Instead of the usual pharmaceuticals, fragrances from licorice and ginger can also block it, as we have observed in experiments. Licorice candies or ginger sticks, as passionate cruisers know, are always offered on the ship after meals and are—especially for children—a good alternative to patches or sleep-inducing pills.

The acetylcholine receptor is also chemically modifiable. It is mainly found on our muscles, and ensures that the muscle reacts with a contraction when the neurotransmitter acetylcholine is released from the nerve endings. We were recently able to show that quinine can block this receptor. Quinine comes from the bark of the cinchona tree and is also present in high concentrations in tonic water. So, anyone who drinks such water in the evening can reduce or even completely stop the increasingly common ailment of nocturnal leg cramps, especially in old age. Rubbing with lavender oil further enhances the effects.

16

When the Nose Goes Blind

> One can go blind or deaf—but to lose the ability to smell has no name of its own. Is it because it's not that bad? Those with a "blind nose" see it differently. They suffer daily from their anosmia. There can be various causes for such a loss of smell. One of them: an infection with the coronavirus.

Those who have lost their sense of smell miss a lot: the scent of lilacs in spring, the sea or freshly mowed meadows, the smell of a loved one, and the wonderful taste of their favorite food or wine. Because those who can no longer smell, can also no longer taste. A whole world without scents. Also, without their own body odor, or rather: without the ability to perceive it. Maybe I stink? Or have bad breath? Some with a "blind nose" wash several times a day or overdose on perfume because they know that their nose can no longer warn them. Not about sweat odor, not about spoiled food, and not when the milk is boiling over. A lost sense of smell often goes hand in hand with a reduced social life, and some of those afflicted even suffer from depression.

The loss of the sense of smell is called anosmia. Since the beginning of the corona pandemic, many people have complained about disturbances in or even the loss of their sense of smell. According to an online survey, 70–80% of those affected experience a temporary restriction, with about two fifths still being unable to smell properly after 2 to 3 months. "The first weeks were the worst," says one affected person. "I couldn't smell anything at all and was really panicking that it would never get better." For some, the same cabbage, the same salad dressing smells different every day; for others, coffee suddenly smells like gasoline.

Researchers see the cause in the fact that the supporting cells, which are responsible for the protection and supply of the olfactory sensory cells, are damaged by the coronavirus. The supporting cells carry a protein called ACE2 on their surface. These proteins are welcome docking stations for the virus's own spike proteins. In this way, the viruses can penetrate the supporting cell and damage it. This, in turn, leads to the individual not being able to smell anything or their normal experience of smelling being disturbed.

The brain suddenly cannot put together the learned scent patterns correctly. We can imagine our 400 olfactory receptors, each responsible for a specific smell, as a scent alphabet. Most scents, such as rose or narcissus, are made up of a mixture of many different scent molecules. Accordingly, many different types of olfactory cells are activated at the same time, creating the "rose pattern" or the "narcissus pattern" in the brain. If some scent letters are missing or suddenly replaced by others, a new pattern is created – or none at all.

Fresh Olfactory Cells Every Month

Anosmia is very rarely congenital and is hardly ever found in young people. When it is, it is usually because obstacles in the nose block the scent molecules on their way to the olfactory cells. These can be polyps or swellings caused by allergies, but also a curvature of the nasal septum. We can imagine our nose as being like a tower, through which the air flows from bottom to top. On the top floor, there is a room with 30 million olfactory cells, the door to which is only slightly open. Therefore, only a small sample can be taken from the air breathed and passed over the olfactory cells. Any swelling or curvature causes the air to be completely bypassed by the olfactory mucosa.

Everyone knows this sensation from a cold: When the common cold, with its adenoviruses, strikes, everything is blocked. No smell penetrates the thick mucus, and the most delicious food is completely tasteless. Usually, the symptoms quickly subside. Also, chemical substances or medications (some antibiotics, almost all cytostatics) can temporarily disable the nose. After discontinuing the therapy, the sense of smell usually returns completely.

It is known that smokers also have a significantly poorer sense of smell because tobacco smoke toxins damage the olfactory cells. Fortunately, the sense of smell normalizes within a few months if one decides to quit smoking.

This miraculous healing takes place because there is a layer of stem cells beneath our olfactory cells that, in humans, regularly and for our entire lives, completely renew all olfactory cells every 4–6 weeks. Thanks to these stem cells, the sense of smell also returns after damage or a cold. However, this process can take up to a year or longer before a complete renewal by the stem cells has occurred. Since the stem cells can renew not only olfactory sensory cells but also supporting cells, the sense of smell also returns completely in most covid patients, although it can take several months or a year for it to happen. This fact has now been determined by a team of ENT doctors in Strasbourg. They tested a group of patients over 12 months. Their tests showed that younger patients and women recovered faster. Regular smell training can also contribute to healing, which can be carried out with various essential oils or spices.

When Orange Juice Smells Like Gasoline

However, if the various viruses manage to infect and kill the stem cells, there is no chance of renewal and, to date, no therapeutic possibility of helping these people. We now know that in about 10% of severe colds, the sense of smell does not return. A chronic inflammation of the paranasal sinuses can also lead to a complete loss of smell.

In the past, anosmia often occurred after car accidents. When the head hits the windshield, the connection of the olfactory cells to the brain is torn. Thanks to airbags, such accidents have become very rare. But even in falls that involve traumatic brain injury or a hard impact to the back of the head, the connection between the nose and brain can be destroyed or the olfactory center can be damaged.

While some people can no longer smell anything, some suffer from a greatly increased sensitivity, even a hypersensitivity, to fragrances (hyperosmia). The intensity of the scents bothers them just as much as insensitivity bothers others. In addition, there are disturbances in the processing of smells in the brain that lead to a completely changed perception of olfactory stimuli. In a case like that, orange juice may suddenly smell like gasoline or solvent—a phenomenon that experts call parosmia. The constant perception of bad, foul smells, so-called cacosmia, represents an especially huge burden on people's quality of life. Such smells may even occur without any fragrance molecules being present, i.e., without olfactory stimulus (phantosmia). This phenomenon is often associated with psychiatric illness.

The Nose Ages Too

The processing centers of fragrance information in the brain can also be affected.

Thus, all neurodegenerative diseases, such as Alzheimer's, ALS and Parkinson's in the final stage, are almost always associated with a complete loss of smell.

In Parkinson's, a reduced sense of smell can be measured 10 years before other symptoms, and thus the smell test can be used as an early diagnostic marker for this severe disease. However, this does not mean that everyone whose sense of smell deteriorates with increasing age automatically suffers from Parkinson's. Conversely, however, those who can still smell well in old age need not worry about having a neurodegenerative disease.

The most common cause of anosmia, or loss of smell, is indeed age. From around the age of 60, we are all not surprised if we need glasses or a hearing aid, but few think about the fact that our sense of smell also deteriorates. Scientific analyses have shown that about 5% of the population suffers from almost complete anosmia, and nearly 20% are affected by a reduced sense of smell (hyposmia). Among those over 75, roughly every third person has almost completely lost their sense of smell. The reason for this is the declining capacity of stem cells to form new olfactory cells. However, we are not helpless in the face of this process. Those who start the right training early can not only improve their sense of smell, but also maintain it longer, while simultaneously engaging in "brain jogging"—more on this in the chapter "From Smell Training to Brain Jogging".

17

Of Stinky Fruits and Moldy Cheese

> Why do we love certain scents and find others disgusting? This is due to our upbringing, but also to personal memories and cultural preferences. Europeans love vanilla and think that dried fish stinks. Asians, on the other hand, appreciate the smell of fish and turn up their noses when they meet Europeans.

We have become accustomed to some scents, while others surprise us out of the blue. Through the direct connection that scents take to the brain, they can quickly evoke memories. Comforting nostalgia or spontaneous disgust—this is not decided by the slow mind, but rather by the quick emotion. The writer Marcel Proust experienced sudden feelings of happiness at the smell of Madeleine cakes, which reminded him of carefree childhood days, while most French people probably ignore Madeleines. The same smell affects different people in very different ways, depending on what they associate with it. While some people happily refuel their cars because the smell of gasoline evokes images of summer holiday trips, Vietnam veterans describe the opposite: They had to douse and burn corpses with gasoline during the war. To this day, they cannot refuel without feeling nauseous and a kind of spontaneous panic.

In addition to personal scent memories, the scents of our culture also play a role, because people within a certain cultural circle gather similar experiences. If mother's milk and baby powder smell like vanilla, it is no wonder that we later find everything that smells like vanilla comforting, soft and delicious. And it is also no surprise that vanilla ice cream is the favorite ice cream of the Germans.

Whether someone finds a stinky fruit worse than a moldy cheese says a lot about their origin. A student from Germany told of when his friend from

Tokyo came to visit: he made sure to clean the refrigerator and throw away the disgustingly stinky blue cheese. "To save me from certain death". Indeed, if you were in charge of running a training camp for Asians, you would have to come up with completely different tests for measuring disgust than you would use for westerners. Being trapped in an elevator with a European after an hour of fitness training, for example. Trapped with a "butter stinker"—a true nightmare! Asians quickly turn up their noses at us because they have far fewer sweat glands and hardly any body hair, hence they also exude a much milder body odor.

Family Scent Connects

Thus, scents can separate people and cultures, but they also connect them. Like the family smell. Various scientific experiments with twins, siblings and their parents suggest that the common genes actually cause a similar body odor. Such experiments are usually carried out using T-shirts that the test subjects wear for several nights. In this way, scent samples are collected that the subjects are then encouraged to sniff. The result was: Family members could identify each other by body odor, and they found this smell more pleasant than that of strangers. The body odor of identical twins is so similar that even trained sniffer dogs can hardly distinguish it.

The scent of the big, wide world is less popular than one might assume. On the contrary: Most people prefer familiar smells. At Christmas, we are pleased by smells like pine branches and cinnamon cookies; in the church, the divine scent of incense conveys a comforting feeling of security and emotional belonging. Everyone also knows that death and the devil stink of sulfur and that strongly scented lilies and chrysanthemums are typical cemetery flowers. Collective scent preferences often also come from the kitchen. A Turk loves the doner kebab haze, the Italian, the scents of pasta and pizza, and the Korean, the unique odor of kimchi, in every sour variant. Some people smell like garlic, which may not be as well received as the stench of cow dung. The latter may remind us of a poor farm with a broken washing machine, but not so in Africa. There, the smell of cow dung is associated with power and prestige. Whoever stinks of it the most is considered the most successful at cattle breeding.

Popular Worldwide: Oranges

Freshness and cleanliness also have very different scent notes. In Germany, a cleaning agent must smell like lemon, so that the housewife can use it with a clear conscience. The Spanish woman trusts only the smell of chlorine, while in Russia, the scent of lilac is considered the cleanest thing in the world. All of these are scent memories from childhood that shape us for life. We pick them up from our environment and from the places where we live.

The smell of the Paris metro is just as characteristic as the famous GDR mix of lignite and disinfectants, which could evoke nostalgic feelings, if it were not associated with such unpleasant memories.

Olfactory city and country walks are now popular all over the world. In Berlin, you can enjoy the scent of linden blossoms, in Singapore, it smells of sharp spices, and in Marseille, of diesel and peppermint. In Bavaria, blooming chestnuts spread the smell of sperm and the "Smell Walk" of New York leads you through clouds of "poison gas" infused with garlic, similar to our river meadows, when the wild garlic blooms in spring.

There are very few scents that are perceived as pleasant or unpleasant in nearly all cultures. People worldwide agree that isovaleric acid absolutely stinks. It reminds us of the smell of cheesy feet. No wonder, because isovaleric acid is actually also found in stinky cheese like Harzer and Romadur. Both cheese and feet activate the same receptor. The most popular scents, on the other hand, not only include the flavoring vanillin, but also fruity scents, like peach, pineapple, orange and green tea. The orange stands for good taste, sweetness and freshness. With other foods, the industry steps in. Food chemists know: ready-made soups and pudding powder must taste different in Asia than in Europe or the USA. Also, tobacco leaves are perfumed according to regional preferences: In the USA, cigarettes are expected to smell like popcorn and barbecue, while the European smoker prefers the smell of fruits and firewood. This explains why scent designers are in demand in all major companies.

18

Get Slim Faster with Bitter Substances

> Brussels sprouts and chicory are not the most popular vegetables, as they taste damn bitter. But bitter substances are healthy. They strengthen the immune system, boost fat burning, and can prevent obesity and diabetes. Moreover, every person tastes differently: How intensely we perceive bitterness is determined by our genes.

It's a bitter truth: What tastes good is not always healthy. Nature had planned it differently, not expecting that there could be an abundance of food. The idea of evolution: Sweet provides a lot of energy, which humans need for life and growth, while bitter should be avoided, often even indicating toxic substances, so that humans and animals will avoid ingesting it. Plants use this to drive away their predators, and thus ensure their survival. Alkaloids such as nicotine, the atropine from the deadly nightshade, and even strychnine, which occurs naturally in the nux vomica, are harmful or even deadly to humans. Additionally, solanine, which occurs in green potatoes, can cause nausea and vomiting.

Our taste sensory cells are a thousand times more sensitive to bitterness than to sweetness, saltiness, or sourness, so as to warn us of the dangers of these toxins. The aversion to bitterness is genetically predetermined and innate. Even newborns will grimace disgustedly at bitterness and spit out whatever they have in their mouths immediately. Even older children can rarely be convinced to try bitter-tasting vegetables. Only over the course of life do we learn to appreciate the benefits of bitter substances.

Bitterness Drives Away the Addiction to Sweetness

For thousands of years, bitter fruits and roots have been used against all sorts of evils and complaints. Hippocrates and Paracelsus reported on the healing powers of bitter substances. The famous abbess Hildegard of Bingen planted bitter herbs in the monastery garden and wrote down two thousand recipes for the use of these herbs in healing. Since 1846, there has been a secret recipe for the production of "Underberg", the most popular of all digestive schnapps, which supposedly contains herbs from 43 countries. The beneficial effect that bitter substances have on the stomach, liver, gallbladder, and pancreas can now also be scientifically proven with the discovery of olfactory receptors in these organs. They prevent gallstones, help against bacteria and viruses, combat the acidification of the body, have an antioxidant effect, and are also the purest fat burners, so that they are recommended for any diet. They direct the fat from the food directly to the place where it is burned—without intermediate storage on the stomach or hip. In addition, bitter foods make you feel full faster; after all, nature wants to protect us from too much bitterness. Ayurvedic teaching praises bitter substances because they are effective against an addiction to sweetness. Everyone who has ever listened to a nutritionist knows what sugar does to the body. Less sugar not only helps protect against diabetes and irritable bowel syndrome, but also promotes dental health, as it causes less tooth decay and gum inflammation. No wonder that there are now a lot of dietary supplements whose producers praise the benefits of natural bitter substances (selling them at a high price).

Super Tasters Get Full Sooner

It is cheaper to consider appropriate types of vegetables when cooking. Many bitter substances are contained in broccoli and cabbage, eggplants, olives, and radicchio. Not to mention the artichoke, which influences the production of bile acid with its bitter substance cynarin, as well as lowering the cholesterol level. Those who want to do even more for their figure and health can supplement their meals with ginger, basil, marjoram, oregano, or rosemary—all full of Bitter! Salads can be nicely complemented with arugula, endive, or dandelion from your own garden. All of these are really bitter and can be briefly blanched if necessary to extract some of the bitter substances before eating. The bitterest natural substance is found in the root of the yellow gentian and is called Amarogentin. Its healing effect was already known to Hildegard of Bingen. A shot glass of it is enough to make the water of 6000 bathtubs taste bitter.

Whether we like a dish or whether we absolutely dislike broccoli or cabbage also depends on how intensely we perceive the bitter components of the food. Although every person has the same 25 receptors for different bitter substances, each of these receptors also has small, inherited differences. They determine whether we are a super taster, a normal taster, or a non-taster. But they also affect whether we particularly like certain types of cabbage or not. Thus, there are entire families who prefer broccoli to kale. These preferences are demonstrably not learned, but innate. In addition, there are people who do not perceive any bitter taste at all. They are at a dietary disadvantage, as they have to eat more than others before their satiety center responds. Super tasters are in the best position, representing about a quarter of the population. They feel full earlier and, on average, weigh about 20% less than people who are insensitive to bitter substances. That bitter substances are good for the figure was proven by another study, involving 500 overweight women and men. They were instructed to continue eating normally, but, for 3 months, they were to take a bitter substance-rich concentrate made from wild herbs. The result: The participants lost an average of four kilos—apparently because they felt full earlier due to the bitter substances.

Once again, organic vegetable cultivation, with many natural varieties, proves to be at an advantage over the modern food industry. It sees the bitter taste as a disruptive factor for sales, and has therefore bred it out of many vegetable varieties to make the vegetables more palatable for the average consumer. A bitter loss!

19

The Scent of Christmas

> Every year, the scent of cookies, gingerbread, and roasted almonds tempts us. Doctors warn against too much sugar, but the spices themselves are old remedies: cinnamon and anise, cardamom and ginger successfully fight infections and pain.

The Three Wise Men came with frankincense and myrrh. Precious, holy scents, no trace of sugar. Today, the Christmas season is undoubtedly an attack on the waistline and the healthy cholesterol level. And the nose is not entirely innocent. Because, when we walk around the Christmas market, as the scents of mulled wine, fir green, and cinnamon stars waft around us, memories are awakened of happy holidays in childhood, the anticipation of gifts and, best of all, a cozy feeling of security among family. The Christmas mood as a nostalgic escape from the cold of winter and everyday life.

But that's not all. Scientists have discovered that the spices associated with Christmas also have a positive effect on those people who are unfamiliar with them, and thus do not associate any memories with them. Even those who have never encountered a potpourri of Christmas spices still succumb to its enchanting effect, because many of the spices that we use for cookies and winter food are healthy, and therefore attractive to humans. Folk medicine has known this for centuries: these spices can stimulate digestion, inhibit inflammation, lower blood sugar, and even relieve pain.

Cookies Instead of Pills

Star anise not only looks beautiful in mulled wine and tastes good in a Christmas tea blend, its antibacterial and antiviral active ingredients also help against winter colds, often also against bronchitis or tonsillitis. The shikimic acid contained in it was even chosen as the base for the flu remedy Tamiflu. If the throat is mucusy or the intestine is blocked, an infusion of the eight-pointed fruit brings quick relief. Anise, which tastes slightly licorice-like due to the essential oil anethole, is also a good remedy for stomach cramps and flatulence. Or you can chew the seed husks. They also help digestion after heavy meals and prevent bad breath. Because of its mild, tasty aroma, anise is often used in pediatric medicine. In addition, there is the expectorant and cough-suppressing effect of ingredients such as anisaldehyde and anisketone. Ground anise is used for delicious cookies, pepper nuts, and Springerle. As an essential oil, anise—similar to clove oil—relieves toothache, but can rarely eliminate its cause.

One indispensable companion through the winter is cloves. They are the flower buds of a more than 10 meter high, exotic tree, which belongs to the family of myrtle plants, and has been known to act as an effective medicine since antiquity. Originally, cloves came from Indonesia, but they are now grown everywhere. They not only give mulled wine and pastries their typical taste, but they also help in digesting the festive red cabbage. Cloves ensure the release of the neurotransmitter serotonin, which makes us happy and stimulates intestinal motility, thus promoting digestion, thanks to the high concentration of eugenol. Interestingly, this works thanks to the clove scent receptor from the nose, which is also found in the intestine.

The antibacterial, analgesic, and anti-inflammatory eugenol can be successfully used against many infections and mycoses. Toothache can be numbed with eugenol by putting a clove in the mouth, and even in obstetrics, clove oil was used for pain relief in the past. Cloves are unbeatable in terms of antioxidants. They protect our cells, prevent the aging process, and strengthen our defenses. The carnation is considered the best radical scavenger among the spices.

The Secret of Gingerbread

One of the oldest spices is cinnamon, which was first used in China and India. The Egyptians also used it for embalming and as incense. The Romans, on the other hand, initially appreciated its effect as an aphrodisiac, before they discovered its delicious taste. For centuries, cinnamon was one of the most expensive spices, bringing the trading nations of Portugal and Holland great wealth. Cinnamon stars contain the essential oils from the cinnamon bark, which are particularly effective at killing harmful germs. At the same time, the cinnamaldehyde goes under the skin: It stimulates the heat receptors of the trigeminal nerve, so that we feel warm around the heart. But beware: do not bake the cinnamon cookies at too hot a temperature, otherwise harmful acrylamide is formed. Also, do not eat too much of it, because the flavoring substance coumarin can be harmful to the liver. Large amounts of coumarin are particularly found in the cheaper Cassia cinnamon, while cinnamon from Ceylon contains only small amounts.

The liver and gallbladder, on the other hand, are especially helped by the essential oils and the pepper-like molecules of cardamom—indispensable for speculoos and Christmas gingerbread, so-called not only because of the ginger used in it, but also because cardamom itself belongs to the ginger family. Cardamom is also one of those spices that have been valued since antiquity. It comes in capsules that protect the aroma of the seeds. Ayurvedic medicine uses cardamom to promote digestion and ignite the life fire. The aromas derived from ginger create sharpness and heat.

The ginger root itself is a very effective and widely used remedy against nausea in Asia. Its ingredient gingerol blocks a receptor that is responsible for most forms of nausea. So, if you have eaten too many cookies, you might want to reach for homemade ginger tea: Simply cut the ginger into small pieces and pour boiling water over it. Ginger also helps against colds in this way! The new German cuisine swears by ginger as a super tuber, and ginger shots are prepared from it: ginger and lemon as concentrated loads of vitamins and minerals to strengthen defenses and fitness and fight against inflammation and general malaise. Optionally, it can be paired with turmeric, to be on the safe side.

Vanilla Against Christmas Stress

Sweet, warm, tempting, simply irresistible: this is the smell of vanilla. When enjoying vanilla crescent cookies or vanilla ice cream, most people are delighted because they remember the enjoyment of mother's milk and many sugary childhood joys. Vanilla lifts the mood, reduces Christmas stress, makes you happy and stimulates digestion.

As for frankincense and myrrh: They are among the oldest known remedies and were therefore already valuable gifts at the time of Jesus' birth, given to him by the Holy Three Kings. As the main ingredients of the famous Egyptian balls "Kyphi," they were chewed to prevent bad breath and inhaled for spiritual rapture. Myrrh also has a disinfecting and wound-healing effect, while its bitter substances calm the intestine. As to the resin of frankincense, it is now known that it contains anti-inflammatory substances, the so-called boswellic acids, which have fewer side effects than corticoids. Frankincense oil can be successfully used against joint complaints and sprains. Not to be underestimated, of course, is also the popular effect of frankincense at the Christmas service: Here, too, memories of distant childhood times are awakened—though, they may, perhaps, be a bit foggy.

20

The Sophisticated Palate Games of Wine

> When wine connoisseurs meet, many opinions will be voiced. Does the noble drop taste of plum or herbs, or does it perhaps smell of cold horse sweat? Mouth, nose and the trigeminal facial nerve participate in the judgment. Whether one considers red wine as an anti-aging drug or joins those who warn against too much alcohol, everyone must decide for themselves.

The 1966 Chateau Haut-Brion from Graves shines deep garnet red in the glass, and the sommelier's nostrils quiver as he smells it. "Stone fruit and raisins, accompanied by cedar wood tones and different spice aromas" is his scent diagnosis. His neighbor, on the other hand, clearly detects plum, flanked by subtle roasting aromas and slightly animalistic tones, while a third expert perceives subtle pipe tobacco, in addition to leather. The impending dispute is settled with a compromise formula: It is a truly great wine, harmonious in its complexity and elegant in the variety of its aromas.

When oenologists, wine merchants and sommeliers meet to taste wines, there is often heated discussion. More vanilla and currant or a hint of resin and fresh wood? The impression of the second wine, a Chateau Palmer from Margaux, is rarely unanimous: It smells intensely of fur and cold horse sweat, the connoisseurs find, accompanied by all sorts of herbs, leather and cedar wood.

The average wine drinker finds it difficult to sense such subtleties because their taste and sense of smell are not well trained. A wine aroma contains more than 140 different fragrances, which accordingly activate many olfactory cell types in our nose, and thus create a complex pattern for "wine" in the brain. Half of them are recognized by everyone immediately, because these are the classic scents that occur in all wines on this planet. The other half are specific to different grape varieties. Only through a lot of practice does one gradually learn the typical pattern for each type of wine, and can then, for example, distinguish a Riesling from a Silvaner. With even more practice, the professionals can identify different growing areas, although their patterns only differ minimally. And sommeliers at the absolute top of their games recognize patterns that only differ in one or two of the more than 80 activated olfactory cell types.

Noble Barrels and Oak Chips

To capture the wine in its full sensory perception, we need to do more than just smell it with the nose. Only deliberate tasting, swirling it in the mouth, provides further information. On the palate, the aromas already perceived through the nose are confirmed and new aroma substances are released by the warmth of the mouth and through chewing. The tongue analyzes the acidity and the sweetness, the salty and the bitter, while the *Nervus trigeminus* detects the alcohol content, as well as, above all, the astringency, the furry saliva-pulling, the typical taste of oak barrel and tannins. If you take a small sip, you also recognize the finish, which occurs when the aromas rise into your nose from the back of the throat.

The grape aromas of young wine usually consist of fresh apple, lemon, peach, currant or floral scent, to which, over the years, aging aromas are added that can taste like raisins, caramel or chocolate. The materials for storage also give off their own notes. If a wine is matured in oak barrels, it takes on the typical barrique taste. Some wood scents are further changed by metabolic processes, resulting in secondary barrique aromas such as coffee, mocha, cinnamon, clove and, above all, vanilla. After 2 years, the expensive oak barrels have released all their flavorings and need to be replaced, which is why wine is often stored in glass or plastic containers and prepared with a bag of oak chips, a method that is now also allowed in Europe and produces a wine that is almost indistinguishable from wine from a wooden barrel, even by connoisseurs.

Valuable Cork

Some wines can taste wonderful without any additives, provided they haven't been spoiled prematurely. You're looking forward to a noble drop, but you can already smell it on the cork: something is not right here. The cork taste is caused by microorganisms in the cork bark that react with a chlorine compound that gets into the cork during washing and bleaching. The end product is called trichloroanisole—a pretty awful taste note. Up to 3% of all wines become undrinkable as a result.

The main supplier of cork is Portugal. That country is home to the cork oak, which can be peeled for the first time at the age of 25. If you then consider that only the third harvest is suitable for cork production and the oak can only be peeled every 9 years, it becomes clear that cork is one of the rarest and most valuable of raw materials. After the harvest, the bark is dried under the open sky for at least half a year, after which it is cooked and disinfected, and, so that its surface is as smooth as possible, coated with a paraffin or silicone wax. With proper storage, it has a lifespan of 10 years or longer. However, the market is seeing greater numbers of cheap corks, with winemakers now increasingly using silicone and glass corks or screw caps. The function of the natural cork is not only to seal the bottles, but also to let the wines "breathe"; in addition, microorganisms are brought into the wine via the cork that demonstrably change the aroma composition of a noble drop for the better. All of this is, of course, not possible with silicone, glass or screw caps.

The Dose Decides

Whether and how much wine is healthy for humans has been a matter of many different opinions throughout the history of medicine. Hippocrates knew 400 years before Christ that, in wine, not only does "veritas" (truth) lie, but also much "sanitas" (health), and thus recommended wine diluted with water for headaches and digestive disorders. The Greeks and Romans used wine as a tonic for convalescents, as a calming and sleeping aid, as a painkiller, and especially as a treatment for many stomach and intestinal diseases. It was used to disinfect wounds, and for compresses, rubs and massages. Modern medicine discovered that wine contained vitamins (C and B6), minerals and trace elements of health-promoting ingredients, especially polyphenols and tannins. Also, the same thing applies to wine as does to most pharmaceuticals:

the dose is decisive, and too much is harmful. Therefore, many experts warn against regular wine consumption, because alcohol is always harmful.

Other scientists, on the other hand, asked themselves years ago: Why do the French, who are known for good food and lavish wine consumption, suffer only half as many heart attacks as other Europeans? Their answer is: regular intake of red wine. This would lower the "bad" LDL cholesterol and promote the "good" HDL cholesterol, as well as reducing the risk of thrombosis. But which specific ingredients give red wine its reputation? The answer is the bioactive phenols. In its skin and seeds, the red grape forms over 100 different types of them, all of which taste different. Since the whole grapes are used for the creation of the rich red color, the health-promoting substances also get into the glass. In the role of classic "free radical scavengers," the phenols protect grapes against bacteria and insects and humans against the evils of age. They are said to reduce the risk of heart attack and diabetes, maintain mental performance and improve sleep. Resveratrol, in particular, has gained fame, not only as an astringent substance on our palate, but also as the purest of anti-aging drugs, one that is supposed to be as effective against skin aging and cancer as it is in favor of good circulation and strong hair growth.

But back to the scent, bouquet, and taste of the grape juice. If it is not so easy for you to distinguish a young Burgundy from a 5-year old Merlot, if you can't identify cherry on the tongue or plum in the finish, but simply recognize whether a wine tastes good to you or not, then be comforted: Even professionals sometimes find it difficult to judge wines. Legendary is the experiment involving ten well-known sommeliers from Parisian gourmet restaurants. They were supposed to taste, in complete darkness, ten different wines, five white and five red. That sounds easy, but none of the professional tongues succeeded in identifying all of the wines. Apparently, the eye "drinks" far more intensely than one might think. Only when the test was reduced to one white and one red did most of them get it right. Incredible, don't you think? Could you have done that too? The test doesn't take long, give it a try!

21

Whether Drugs or Truffles: Dogs are Perfect Sniffers

> Scents, smells, and stench are its world, and it can follow even the finest traces: the dog is an absolute top nose. Dogs not only find drugs in flight luggage and truffles under trees, but also successfully search for rare animal species and pests in the forest.

Of course, a dog immediately recognizes its owner by their scent. Not only that: it also knows how they are feeling. Joy, fear, and stress emit certain scent signals that the dog perceives. An anxious postal worker can indeed provoke different reactions. Some dogs just bark at them because the stranger, who seems inferior due to their fear, needs to kindly exit their territory. More aggressive counterparts feel provoked by the foreign smell, and thus bite. Confident dogs, on the other hand, often react quite calmly.

But even dogs are not born with a perfect sense of smell; they must first develop it. With targeted training, a dog can be taught a great deal of scent information that it can use to provide good service to humans. At airports, sniffer dogs are used to find drugs or explosives. If a person is missing, the sniffer dog can follow them, even after days have passed. In avalanches or earthquakes, trained dogs can locate buried people and save lives. In the medical field, they are now recognized experts in many areas. More on this in the next chapter.

Sniffer dogs are becoming increasingly important in such diverse areas as safety code adherence and agricultural use, now even branching out to species protection. A well-trained dog nose can find various types of fungi, especially mold, even in hard-to-reach or hidden places in a building. In other studies from Africa and Latin America, dogs were able to track down not only bedbugs, but also the dangerous assassin bugs that transmit the life-threatening Chagas disease, from which more than 1.5 million people suffer in Argentina. According to the WHO, more than 10,000 people die from it each year. The bugs hide in cracks and room niches. For humans, they are very difficult to find, but sniffer dogs have no problem. In agriculture and forestry, studies are underway to train sniffer dogs to detect pests. Beetles and their larvae, which can multiply wildly in the increasingly warmer and drier climate in sick trees, are especially the focus of use. This is particularly true for the Asian long-horned beetle, a dangerous wood pest and neozoon.

On the Trail of Rare Species

So-called species sniffer dogs are becoming increasingly important. As studies on otters in the Upper Lusatian heath and pond landscape have shown, these dogs can not only track down sought-after animals, but also distinguish very closely related species based on their feces with high selectivity and specificity. Their success rate can be up to 100% (a similar experiment with human experts only achieved an accuracy of roughly 70%), and they have even been known to recognize individual animals, a fact that is of particularly great importance for nature conservation.

The search for endangered amphibian species, such as newts, toads, and reptiles, is also at the center of interest for conservationists. Water-dwelling amphibians in particular spend a lot of time on land in well-protected hiding places. In a pilot project, dogs discovered not only various species of newts, but also the rare natterjack toad, even in the densest habitat. The successful use of species sniffer dogs in locating hidden-living species has increased significantly worldwide. An overview article in the journal "Methods in Ecology and Evolution" by the Helmholtz Institute for Environmental Research, written together with the Institute for Zoo and Wildlife Research in Berlin, summarized the large number of scientific publications on this subject in 2022 and showed that data has been collected on successful search dogs for over 400 animal species and even about 50 plant species from 60 countries.

Dogs as Truffle Hunters

Some dogs are more active in the leisure sector, like the dog of a truffle hunter named Franco in the Italian Piedmont. A travel group, led by Franco, set out one evening for the gnarled old oaks standing at the edge of his vineyards.

However, no one looked at the beautiful landscape, with its gentle hills and snow-covered mountains in the distance. All participants of our group were too busy looking intently at the ground: Were there any signs yet of the "white gold of Piedmont": the truffle? Very unlikely, because the coveted fungus, which does not form chlorophyll itself, lives in symbiosis with other plants up to 40 cm underground, mostly in the root network of chestnuts and oaks. There, the truffle quietly stinks, unnoticed by human noses. Skatole, indole and dimethyl sulfite, typical fecal odors, as well as the steroid androstenol, the sexual pheromone of the boar, are its characteristic scent marks. Pigs love this pheromone, which is why it is mostly pigs that are used for truffle hunting in the French Périgord. However, the truffle hunter must be very careful that they do not eat the mushrooms themselves out of sheer enthusiasm.

Franco prefers to trust his dog. The dog immediately starts sniffing around frantically, picking up a trail. When he stops, Franco begins to dig. Everyone is somewhat disappointed when, shortly thereafter, he brings an ugly beige-gray lump to the surface. Small and shriveled. Is this supposed to be the famous truffle that everyone raves about? For which they spend thousands of euros at the markets in Piedmont? Franco's dog probably finds it strange too. Disinterested, he turns away after receiving his treat.

Training a truffle dog takes a whole year. This process could go faster in the future if electronic noses are used. Scientists and truffle hunters are experimenting with various techniques to simplify the search. In Switzerland, such electronic noses are already being used to check whether a boar has been successfully castrated. If not, it will later smell and taste of androstenol, skatole, or indole. Its meat would be inedible.

22

Detection Dogs in Medical Use

> Will dogs replace medical tests and examinations in the future? Even today, it is known that they can recognize changes in body odors, and can thus sniff out diseases like diabetes, cancer, and even COVID. New studies show that their noses sometimes work better than conventional tests, even with Long COVID.

Due to their excellent sense of smell, dogs are able to detect changes in body odor that are based on both psychological and physical changes, and that can be of medical significance. It has long been scientifically proven that it is easy for dogs to recognize and distinguish human emotional states. Thus, dogs were able to reliably distinguish visitors of cinemas who had watched either a romance, a horror film, or a pornographic movie.

Dogs accompany epileptics and diabetics, serving as four-legged alarm systems, because they can sniff out the signs of an impending seizure and warn the patients in time (see also Chap. 11). Whether they recognize the metabolic changes in body odor or the stress reactions of the body preceding a seizure is unclear.

Dogs are even used as cancer specialists, because it has been found that they can identify certain tumors very early and accurately. Bladder and lung cancer, for example, but also breast, skin, or colon cancer. It is quite possible that we will soon be spared the bothersome procedure currently needed to screen for colon cancer.

Even during the pandemic, it became apparent how helpful the use of a dog's nose can be. Scientific studies have shown that detection dogs can be trained to recognize COVID-positive patients at an early stage, with high probability. Corona detection dogs were successfully used at airports in

Helsinki and Dubai. A scientific study at the University of Veterinary Medicine Hannover in 2020 showed that trained dogs could recognize the genetic material of SARS-CoV-2 pathogens, as well as urine and sweat from COVID-19-positive patients. And with an accuracy of over 90%. They are thus better than conventional rapid tests. Dogs can also be used for real-time screening, e.g., for stadium visitors or at various events and museum visits. Currently, efforts are being made to extend these studies to other pathogens, with initial positive results having already become available for various influenza and HPV viruses.

Interestingly, there are new studies from 2023 in which dogs also sniffed out Long COVID patients whose PCR tests had consistently been negative. A research team from the University of Veterinary Medicine Hannover published their results in 2022. According to them, dogs could still recognize the previous illness in the exhaled breath of the patients, even though the clinical tests (antibody detection) were negative. The trained dogs do not smell the viruses themselves, but rather small volatile molecules that arise from the altered metabolism during viral infections. Thus, they are able to distinguish acute COVID infections from Long COVID, without even needing to smell the patient directly, relying only on saliva or sweat samples.

Initial data even suggest that they can recognize not only different corona strains with over 80% sensitivity, but also different virus strains in other respiratory infections.

Will Corona detection dogs be the COVID detectors, or the "Medical Detectives," of the future? In England, detection dogs are already roaming through crowds to detect certain diseases. Such live screening could possibly prevent some closures.

23

Animal Super Noses Help Humans

> When their noses are properly trained, dogs, but also rats, wasps, and even birds, can provide valuable services to humans. They are ideal mine detectors, diligent explosives experts, and cost-effective garbage collectors.

It's hard to believe, but rats can also be lifesavers. Admittedly, they once brought the plague to humans, which resulted in a persistently bad image. Their presence in great numbers is also rarely welcomed. But humans tend to underestimate the animals, as rats are smart and teachable. Moreover, they are light in weight. All qualities that make them ideal mine detectors. Their low weight will not trigger a detonation mechanism, and they learn much faster than dogs. While a dog takes a year to find mines, the rat only needs four months.

Two of the star members of their species, Lola and Espejo, became world-famous. They were part of the first mine-detecting rat squadron in Colombia, whose fields were completely mined and largely unusable after five years of civil war. Lola and Espejo quickly learned to sniff out the TNT packets and were rewarded with their fair wage: heaps of crackers. This made them much more cost-effective than professional mine detection devices, which are often unaffordable for the affected countries. In Africa, people often risk their lives going mine hunting. Honeybees have also been used for this purpose. In experiments, they flew over the fields to collect nectar, returning afterwards with tiny TNT particles. Or not. Many bees simply flew away without caring about their scientific mission.

Experiments with wasps worked better. Biologist Glen Rains from the University of Georgia was one of the first to recognize their potential. When he explained to anti-terror experts at the US Department of Defense that his team was working on a multipurpose weapon that was cost-effective and self-replicating, everyone was thrilled. However, when the researcher revealed that it was wasps he was training, the experts were appalled. Rains argued that the wasps were capable of detecting mustard gas, anthrax, and even sarin, all explosives that are undetectable by humans. Despite the early skepticism, the military became so fascinated by the idea that they eventually financially supported the project.

How do you train wasps for such missions? Quite simply: with sugar water and syrup. "You can easily condition thousands of wasps to [detect] a specific substance within a few minutes," explains a biologist from the team. And the success rate is higher than that of trained sniffer dogs. With the help of a detector, about the size of a tin can, and a camera, the animals and their reactions can be observed. In this "Wasp Hound" experiment, the wasps return after each flight and deliver their information. The results are forwarded to a computer.

The sense of smell in birds was long underestimated. In fact, it was disputed whether they could smell at all. It is now known that their sense of smell is actually very pronounced. Birds have an extensive repertoire of olfactory receptors, 550 different ones, almost 200 more than we humans. They can recognize and distinguish a wide variety of smells, even in very low concentrations. However, the perception of smell in birds is more variable than in other mammals. Some species have developed an excellent sense of smell, while it is less pronounced in others. For example, corvids like jackdaws, crows, and ravens use their sense of smell intensively, as do seabirds, which need it to find their food. Conversely, it is less important for garden birds. There are now well-founded scientific studies confirming that migratory birds use their sense of smell for navigation. Even at great heights, they can still follow the scent trail on land.

Both corvids and pigeons have demonstrated scientifically above-average cognitive abilities. They are among the most intelligent animals of all. They use tools, can find food hidden under leaves or in containers, and solve puzzles. In the Swedish city of Södertalje, they are expected to ensure clean streets in the future. The start-up "Corvid Cleaning" wants to encourage the birds to participate in cleaning the streets and the city through rewards. With their good sense of smell and cognitive abilities, corvids are ideal for this purpose. It is no problem for them to identify different waste components by smell, even from a greater height. The core of the idea is a machine that dispenses food as a reward for each piece of waste or cigarette butt inserted into it. Hooded crows and jackdaws are particularly common for use in this endeavor, and they learn very quickly. Since these animals constantly observe their conspecifics and learn to imitate everything, they do not even need to be specially trained for the job. They could soon be on the move as an excellent feathered city cleaning crew, significantly reducing the city's cleaning costs.

The model for the Swedes was possibly the western French amusement park "PUY DU FOU." There, crows have been used as cleaners for years. The park now hosts six trained crow families, all with their beaks full of work. In various Dutch cities, there are now the so-called CROW Bars—containers into which crows can throw collected cigarette butts. If the bar recognizes the item, it spits out a peanut in return. However, it is still unclear to what extent collecting waste is hazardous to the birds' health.

Animals in service to humans are cost-effective, but often unpredictable. The dog is moody and refuses to work, the bees buzz off to the next area, and the wasps disappear never to return to their tin cans. Machines are, of course, more precise and reliable. And who knows: maybe soon we will be able to use the olfactory receptors without the animals—like with the artificial truffle nose

24

Divine Fragrances and Worldly Perfumes

> Incense was considered the scent of the gods in ancient Egypt and was used for the embalming of the divine Pharaoh. Herbs, resins, and oils were valuable gifts of fragrance. The Quran praises camphor and musk, while Judith from the Old Testament and the wise and beautiful Cleopatra used the fragrances of cinnamon, jasmine, and rose petals, mainly for political purposes.

The fragrances of plants and herbs and the allure of scents have been known to humans for thousands of years. Already in the Neolithic period, ceramic vessels were being crafted for the storage of resins. In Mesopotamia, temples were built in the 3rd to 1st millennia BC, in whose walls small channels led to caves from which the soothing scent of gardenia flowed into the house of God. Spices and herbs were extremely precious, just good enough for the gods, who were to be appeased with them.

Noah and the other survivors burned cedar wood and myrtle as a sign of their gratitude after their rescue from the flood. The Three Wise Men brought the baby Jesus frankincense and myrrh as valuable gifts. Incense is still considered by the Catholic Church today as a symbol of purification and veneration, as well as a sign of the presence of the Holy Spirit. In ancient Egypt, one could recognize the presence of a god by it, even before they showed themselves in their divine form. Incense vessels and inscriptions testify to its importance, for example, in the birth cycle of Ramses II, 1279 BC: "Your smell pleases me, your scent is that of the land of the gods, your fragrance is that of incense."

The divine Pharaoh was accompanied by the scent of incense from birth to death, and it preserved him from decay during embalming. Since incense was considered the inherent scent of the gods, the deceased was thus olfactorily included in their community. The gift of incense from the Three Wise Men to the baby Jesus is also a reference to his divinity. Myrrh, on the other hand, is usually interpreted as a reference to burial, because the plant is one of the classic anointing aromas.

When incense is dispersed in a church, it is warmer than the surrounding air and floats upwards into heavenly realms, a symbol for the ascent of thoughts to God. Its scent not only has a calming effect, but is also slightly intoxicating. Catholic believers are familiar with it from childhood, and it gives them a feeling of home and belonging. And because the scent of incense also clings to clothing, they carry the scent message out into the world after mass. The Catholic Church can boast of having developed an unmistakable corporate scent and having invented the marketing of a scent associated with modern advertising agencies, this as early the birth of Christ, a practice that has since been perfected.

The Egyptians used not only incense for the embalming of their dead, but also other aromatic resins and oils. And they burned fragrances in honor of the sun god Ra: resins and plant essences at sunrise, myrrh and the sap of the balsam tree at the zenith, and all sorts of refined mixtures at sunset. Also "per fumun"—through fragrant smoke, whose recipe only the priests knew—the Romans conveyed their requests to the gods, and thus gave perfume its name. Whether it was the Etruscans or the Sumerians, the Egyptians, Greeks, Chinese, Persians or Hebrews, they all used fragrant substances from nature that they stored in pots and jars, as we can still see today on frescoes and wall panels.

In Islam, good smells promise heavenly joys: "The pious drink in paradise from a cup into which camphor is mixed", it says in the Quran, verse 76.5, namely, wine, "which is sealed with musk". In the twelfth surah, "Joseph", the title character's jealous brothers try to convince the old father that his son is dead. But Joseph—having come into wealth and prestige in Egypt—sends his father his shirt as a sign of life. From afar, the father perceives the smell of his son, and when they lay the shirt on his face, his blindness is healed (verses 92–96).

Romance with Rose Petals

Seductive scents could also be used very profitably for political purposes. As early as Old Testament times, as the Bible describes, the adulteress Judith conquered the young man Holofernes with fragrances: "I have sprinkled my bed with myrrh, aloe and cinnamon, come let us make love." And her success is legendary: The Jewish people were able to free themselves from the oppression of the Assyrians.

The beautiful and clever Cleopatra is also said to have prepared for her romance with the Roman general Marc Antony with intoxicating floral scents. In this case: with a profusion of rose petals. The entire floor of her room was strewn with them, according to tradition, and she had also anointed her body with a mixture of jasmine oil, rose oil, and honey. The queen could choose from 200 scents at that time. Caravans crossed mountains and deserts, ships crossed the world's oceans to deliver the valuable raw materials. Those who could afford to smell good undoubtedly had money, power, and prestige.

The Arabs are considered the inventors of distillation. Perfumes as we know them today have been produced since the 14th century, a mixture of essential oils and alcohol, which was often used as a remedy at the same time. The ingredients often came from far away, arriving with the trading ships coming from the Orient to Europe. The wealth of cities and homelands being thus founded. The center of sea trade at the time was the Republic of Venice, whose wealthy citizens soon developed a sense for the refinement of manners. Catherine de Medici, who married Henry II in 1533, brought perfumery from her hometown of Florence to France. But it was not until the reign of the Sun King Louis XIV that interest in fragrances skyrocketed, as he was highly sensitive to smells and had his perfumer mix a different scent for him every day to combat the often coarse "vapors" of the palace and the miasmas of the environment. Although soap production gradually increased, it was still believed that water would drain the spirits from the body, and it was thus used sparingly. They would try to mask the resulting uncleanliness with fragrances and, instead of water, scented water—the Eau de Toilette—was used for daily hygiene, a substance that still carries its scandalous name today.

World Success from Glockengasse

The perfume consumption of the royal mistress Madame de Pompadour, who earned merit at the court of Louis XV for promoting music, theater, and a civilized lifestyle, is legendary. Under her influence, an entire industry of luxury goods was created that, in addition to perfume production, was primarily concerned with the production of perfumed gloves. Her own expenses for perfume and fragrant essences were apparently so high that the royal finance minister had great difficulty accommodating them in the budget. In England, even the morally strict Queen Elizabeth I indulged in the luxury of precious fragrances and cosmetics. Like Pompadour, she was fond of perfumed gloves, wore decorated vessels around her neck with the stimulating aromas of cinnamon and cloves, and had the entire palace, including the wallpapers and furniture, scented. It was the Protestants who stopped these excessive sensory pleasures, issuing a decree according to which virgins and widows were accused of witchcraft as soon as they tricked someone into marriage using perfumes, false hair, or other "unfair" means.

The secret recipe for the most famous German fragrance ever mixed comes from the Italian immigrant family Farina: they called it Eau de Cologne, after the German city in which they lived, in the house of the well-known merchant Wilhelm Mühlens, at Glockengasse 4711. SThere, their eponymous water, made from citrus oils, bergamot, cedar, grapefruit, and various herbs, started a sensational triumph around the world in the 18th century. Its clean and fresh smell convinced people of its healthy and invigorating quality. Export flourished early on, as the manufacturer's promises proved full-bodied. The water was supposed to work not only as a fragrance, but also as a remedy: "It is a wonderful antidote against all sorts of poison and an excellent preservative against the plague…jaundice, catarrh, fainting, foul breath…dispels colic and soothes stomach pain, dissolves side stitches and chest diseases, which arise from rising winds and cold feet…is excellent against toothache…"

The most famous fan of the Cologne water was Napoleon Bonaparte, who may also have hoped for relief from his stomach ailment. He allegedly used up to 60 liters a month, with his entire surroundings, including his horse, being perfumed with it.

By the end of the 19th century, the reputation of the legendary water had even reached Turkey. The personal physician of Sultan Abdülhamid II had it imported on a large scale, as he found that the alcoholic Eau de Cologne could be effectively used against bacteria. Eventually, the Sultan agreed to the establishment of a perfume factory in the Ottoman Empire itself. Since 1882, the healing water has been produced inexpensively under the name Kolonya and is appreciated by citizens from all walks of life.

25

In the Frenzy of Fragrances

> A perfume always lends us a touch of luxury. Even if fragrance preferences change like fashions, they are an expression of the individual, as well as being a mirror of society. The latest fragrance creations contain active scents based on scientific research, because we have, indeed, found scents that have a predictable effect on people.

The center of perfume production in the 19th century was the southern French city of Grasse, where fragrance plants had been growing for centuries. Tanners had lived there since the Middle Ages, practicing the odorous trade of leather processing. But then, perfumed gloves came into fashion, and the tanners first transformed into glove perfumers, before eventually dedicating themselves entirely to perfume production, as was the case for Jean de Galimard. The companies Galimard, Molinard, and Fragonard are still famous addresses in the "world city of fragrances". In the local museum, you can watch how the precious oils are obtained: through distillation, through extraction of the fragrances, or even through the use of fat as a fragrance store. This latter method—the oldest method there is, vividly described in the book "Perfume"—is, however, very time-consuming and expensive, and is therefore hardly ever used today. The basic components of a perfume are 80 percent alcohol, mixed with distilled water and dissolved fragrance essences.

Valuable raw materials such as flowers, herbs, resins, and woods are still in use. One of the most noble and expensive essential oils is rose oil. Roses contain so little oil that five tons of rose petals are needed to obtain one liter of essential oil. Animal scent notes can also be valuable, even if their origin sometimes seems rather unappetizing. For example, the ambergris popular in men's fragrances is actually the dried regurgitation of a sperm whale. In other words, whale vomit—large gray lumps that float in the sea. The animals expel the indigestible remains of their food in this way, after which it dries up to become an extremely precious raw material. Thus, a group of fishermen in Yemen was overjoyed to find a whopping 127 kilos of ambergris in the belly of a dead sperm whale in the spring of 2021. Considering the average wage of a fisherman of 800 euros annually and the kilo price of ambergris, which is between 10,000 and 30,000 euros, you know that the fishermen and their families are now set for life, with enough money left over to also care for the needy in the village.

Until the mid-20th century, perfumes remained a luxury item. Only through the production and distribution of synthetic fragrances did they become cheaper. In this process, fragrances are analyzed and copied, but completely new compositions are also created. Synthetic fragrances have the advantage that they can be used as intensively as desired. But they are always only individual fragrance components from the complex mixture of a natural essential oil. The first fragrance with synthetic components was the famous Chanel No.5 from 1921—composed of 80 ingredients and as timeless as Coco Chanel's Little Black Dress. She herself praised it as the "scent of a Nordic morning by the lake". Even film star Marilyn Monroe appreciated its fragrance, over 30 years later: Instead of pajamas, the beauty let the interested public know, she wore nothing at night but a few drops of Chanel No.5 on her body. To this day, Chanel No.5, with its floral-fruity notes, is considered a classic. However, it no longer smells exactly as it did then, but has rather been carefully adapted to the fragrance trends of the new era.

The Personal Scent Decides

Today, perfumes consist of a multitude of different fragrances, some made up of more than a hundred different components. Approximately 200 natural oils of plant or animal origin are supplemented by over 2000 synthetic products, so that the perfumer has a wide range of fragrance notes available for composition, which can also take into account ecological and health aspects such as allergies. There are four fragrance classes: the standard perfume, with a proportion of 15–30 percent of so-called fragrances, the Eau de Parfum, with up to 14 percent fragrances, the Eau de Toilette, with between six and nine percent fragrances, and finally, the Eau de Cologne, with the strongest dilution and only up to five percent fragrances.

In the composition, the perfumers pay attention to a balanced ratio of top, heart, and base notes. The top note should be interesting and arouse curiosity, but it evaporates quickly. Bergamot and citrus notes are mainly used here. Then follows the heart note, the character of the perfume, usually made from floral aromas that can present strongly for several hours. The base note should last a long time and can contain woody or animalistic scents, such as musk. Often, it also smells like moss, earth, or leather. The individual scent only unfolds on the customer's skin, as it depends on their own scent, the fat content of the skin, and the composition of the microorganisms, which is different for every person. Therefore, it is advisable to try out a perfume for a few hours and not to be tempted to make a spontaneous purchase in a duty-free shop or drugstore.

The temptation to make such a perfume purchase is great, because, as in fashion, trends also change in fragrances. Heavy sensual notes like "Opium" by Yves Saint Laurent and "Poison" by Dior came out when the AIDS virus threatened people and it was thought that one should preferably live monogamously. "Cocooning" came with light, discreet scents, and Calvin Klein's "CK One," with its unisex lifestyle message, became one of the most popular perfumes of the late 20th century. The so-called pheromone perfumes, scents with sexual attractants, which are attributed similar irresistible powers as the pheromones of animals, are also supposed to be suitable for both sexes. However, there is not a single scientific proof testifying to their effect. Seduction is not quite that simple.

Effective Scents for Relaxation and Trust

In contrast, our laboratory at the Ruhr University Bochum has published a number of scientific proofs concerning scent effects in humans, including a number of predictable effects that we were able to identify. These effective scents can stimulate and activate or calm and relax. Others promote communication, trust, and affection. Therefore, it is not surprising that perfumes are now being created that contain these effective scents. The first creation of such a perfume was "Knowledge", a scent for the 50th anniversary of the Ruhr University. The well-known Berlin perfumer Geza Schön composed the perfume from over 40 effective scents that were published by our laboratory. "Knowledge" is supposed to promote activity, attention, and performance in its top note, while, at the same time, have a relaxing effect through the heart note and ensuring good communication and trust among people in the base note.

A similar principle is applied in a perfume series from the company "Amatrius", for which Geza Schön has divided various effective scents into the four perfumes "Enjoy Me", "Love Me", "Unplug Me" and "Recharge Me". Depending on what the wearer—male or female—hopes to accomplish, he or she can achieve the desired effect with the right perfume. Geza Schön was also the first perfumer to fully embrace the minimalist approach to perfume-making: the molecule perfumes. While classic perfumes offer a mix of many fragrance notes, they consist of a single scent that is completely produced in the laboratory. No more wasting tons of roses; bring on the synthetic imitations Geraniol, Citronellol, and 2-phenylethanol. Resource-saving, clean, unisex, and suitable for fashion, design and the sustainable lifestyle of Generation Y. Moreover, less irritating for allergy sufferers than many natural products. Geza Schön's "Molecule 01", for example, consists of a single molecule, the "Iso E Super Gamma", which emits a velvety, woody scent and smells like musk or fresh laundry. The same applies to "Molecule 02", which contains the scent Hedion. For both fragrances, our laboratory was able to prove that they activate a human pheromone receptor and influence trust and communication, as well as reciprocity, the act of mutual exchange. However, as long as the substance is still in the bottle, it does not yet constitute the actual scent. This is because it only develops on the skin of the user, different for every person, completely individual and unique.

Which conventional fragrances will shape the year 2022 seems to have not yet been decided among the experts of women's magazines. Are they the functional fragrances or the molecule perfumes? Some suspect that fragrances need to radiate strength. "Si Intense" by Giorgio Armani perhaps? With its surfeit of sensuality and self-confidence. Or perhaps something rather romantic, rich and floral like "Carolina Herrera For Women"? A fragrance with jasmine, night hyacinth, sandalwood, amber and musk. Definitely magnificent and feminine, preferably with leathery undertones or earthy nuances. "For a feeling of security", says Isabelle Abram, perfumer at Givaudan, explaining why she composed her new Nivea perfume: a mixture of fresh laundry and the typical scent of skin cream. A hint of freshness, a freshly made bed and clean clothes—what more can a person want in a world full of problems?

26

Marketing with a Feel-Good Factor

> Scents can create a very special atmosphere. This is a fact that is also known by the professionals of scent agencies, whose job it is to ensure that their clients' customers' buying mood will increase. Or the satisfaction of their hotel guests. Or the happiness of their airline passengers. "Neuromarketing" is the strategy, and scents are one of its most successful tools.

An industry breathes a sigh of relief: Finally, customers can shop again without a mask. No fogged glasses, no crumpled ears, and no grumpy rushing to get everything necessary done quickly. Instead, relaxed faces with red-painted lips and curious noses that had to hide behind fleece for so long. People are strolling through the shops again, being seduced into consuming by soft music and pleasant scents. This is the mood that is hoped will be encouraged in the department store and the supermarket. So: reduce the pace and interest the customer in the offer with targeted stimuli. Only those who take their time, not simply working their way through their shopping list, will see which other products are on the shelves.

The first hurdle is usually found in the entrance area of the supermarket - the so-called pre-cash area. There, a bakery entices with fresh bread, rolls that come from the baking machine on site, and the finest cakes. How good they smell! The customer may have already smelled it on the street, as some bakeries spread artificial bread scents in the pedestrian zones. Now, the potential buyers can convince themselves: The offer really looks extremely tempting. Already, he or she has a slight appetite and forgets that he or she actually only wanted to buy the essentials. This suits the supermarket, as customers with a feeling of hunger buy significantly more. Even food they don't actually need.

To increase the desire to buy, tricks from the chemistry box are also used: The bakeries use artificial additives and, in the fruit and vegetable department, the oranges are sprayed with orange scent. One supermarket chain had the most success with a scent mix of rosewood, orange and lavender, which was sprayed from the ceiling to put the customers in a comfortable mood. The industry paper "Lebensmittelzeitung" reported an increase in sales of 40%.

Scent Landscapes for Women's Fashion

Scent devices are often used in problem areas. If it smells slightly musty at the bottle return or at the service counter for fish and cheese, the devices ensure that the unpleasant smells are destroyed and replaced by an herbal scent that is perceived as pleasant. Professionals therefore first carry out an olfactory room analysis. Then, the air is cleaned and subsequently enriched with a sophisticated scent composition. Very different scents are used, depending on the location. Department stores usually work with a "multiple approach". They offer several scents for the different customer noses in individual departments, and also may create entire scent landscapes: spring flowers for women's fashion, fresh sea breeze for summer outfits, and cinnamon and caramel for Christmas so as to stimulate the sale of gifts.

The customer must be imagined as a happy victim: He reacts as he should, and while he doesn't know why, he is highly satisfied with his purchase decision. Often, he has not even noticed the scent that has allowed him to become so relaxed as he engages in consumption, because it is dosed in such a way that it creates a pleasant atmosphere, but is not intrusive. Only when the concentration of fragrances exceeds a certain threshold can the scent be smelled at all. Experts speak of the perception threshold. At an even higher concentration, the so-called recognition threshold, it can be identified. And, finally, at the difference threshold, it can be distinguished from other scents based on its intensity. From level to level, about ten times the dose is needed. But professional scenters usually don't let it go that far. After all, the scent should not be intrusive or, even worse, get on the visitor's nerves. A deterrent example was provided by the American label "Abercrombie & Fitch", whose stores puffed scent clouds into the neighborhood of such intensity that there were complaints.

The Special Scent of a New Car

And there is another thing to consider: The scent must match the product well. When people smell lilies, a study found, they buy more flower fertilizer, but not necessarily more mineral water. Now, one might ask: Why should flower fertilizer smell like lilies? It probably has something to do with flowers that are already blooming in the customer's mind's eye. At any rate, one soon gets used to it, and only buys this fragrant fertilizer and no other. Experts therefore recommend delivering a matching scent with the launch of a new product. But it should be a scent that is already "brand-congruent", so that the customer immediately has "brand-specific associations".

One very special type of brand is the new car. Whether it's a Smart, an Opel or a Jaguar: The customer has chosen it with all their heart and paid a lot of money, so the good piece should also smell expensive for a long time. What most buyers don't know: The new car scent is actually made up of a lot of unwanted inherent smells from production, which are more or less skillfully masked with fragrances. The steadily increasing number of plastic components, which make the new car lighter than its predecessor, do not themselves always smell particularly attractively. Added to this are leather, rubber and fabric fumes enriched with tanning smells and solvents, plasticizers, paints, adhesive and foam materials. All in all, a mixture that stinks rather than smells.

And this is where the criticism begins: The scent design uses nasty tricks to deceive and manipulate the customer. In particular, this "masking function" of some scent compositions is pure pretense; after all, obvious defects are cleverly covered up. Not to mention that, now, to the visual and acoustic attacks of advertising is added the particularly subtle form of olfactory seduction. Other critics emphasize that the unnecessary use of fragrances can lead to more allergies, headaches and other complaints.

The victim, in this case, the car buyer, usually feels neither sick nor deceived. The purchase of a new car is an extremely positive emotional experience, so its scent is also enthusiastically welcomed. And if, one day, jogging shoes, wet towels or even cigarettes have polluted the air, the car owner is happy to reach for the "New Car" brand spray to restore the coveted scent.

Or she hangs a magic tree on the mirror. Although it looks like a cardboard fir, it doesn't actually smell like pine or spruce, but usually like vanilla. Or just like "New Car".

Corporate Scent: A Touch of Well-Being

A touch of luxury - that is the key to success. Airlines and hotels also take advantage of this. One set of pioneers was the managers of "Singapore Airlines", who have been sending their identically cordial and well-dressed flight attendants to provide passengers with scented hot towels since the 1990s. "Stefan Floridian Waters" was one of the first corporate scents, a mix of rose, lavender and citrus fruits, which forever associates the "Singapore Airlines" brand with relaxation, well-being and safe comfort for passengers. Singapore was one of the pioneers in scent marketing. In Asia, entire shopping centers had already been made fragrant, while in Europe, the windows were still being dutifully opened for ventilation.

Gradually, international hotel chains followed suit, as modern managers discovered that the first ten minutes are crucial for the guest's judgment. Colors, sounds, and, indeed, a pleasant scent should immediately create a good impression in the lobby. For example, the "Westin", which belongs to the "Starwood" hotel chain, introduced one of the first hotel-owned fragrance lines with the scent "White Tea". White tea, ivy, geraniums, and freesias greet the guests, while competition from the "Sheraton" wafted in a little later in the form of a hint of "Welcoming Warmth": fig, bergamot, jasmine, and freesias. In the "Shangri-La", on the other hand, you will be greeted by the sweet and seductive smell of vanilla, sandalwood, and musk. Of course, you can also buy these scents and take the pleasant atmosphere home in the form of soap, shampoo, or creams. Until next time!

Easily recognizable, unmistakable brand scents are now as much a part of the corporate identity when introducing new brands as logos, colors, corporate culture, or appearance. However, scent designers have a problem, and that is the customer. What does she like to smell at all? Which mix is the olfactory equivalent to the pleasing elevator music that everyone likes, or at least tolerates without complaint? No one knows exactly, but there are now a number of studies that at least say: A pleasant scent mix that matches the range of products for sale encourages the customer to linger longer and consume more. A floral scent fits into the garden center, the fruit and vegetable department of the supermarket increases its sales with strawberry scent, and the travel agency benefits from the scent of coconuts. If there are also other stimuli, such as photos of palm beaches and soft music, the customer is completely convinced. He feels happy and not at all manipulated. Which, of course, is a complete misconception.

27

From Smell Training to Brain Jogging

> The cold came and the sense of smell went. Due to illness or simply aging, humans' sense of smell can diminish. In many cases, however, smell training can help. Daily practice can not only bring back the smells, but also stimulate the brain to perform better.

With the arrival of autumn comes the common cold, and with it, suddenly, the familiar experience: everything tastes like nothing. The nose is blocked, and neither the scent of your favorite food nor your own perfume can penetrate the mucus. This is normal. Only if the condition persists should one consider having one's sense of smell tested. A doctor can identify where the causes lie. Is it in the olfactory system in the nose, because the cold viruses have permanently damaged the olfactory cells? Or in the processing of smells in the brain, because a neurodegenerative disease is looming?

For the smell test, there are special smell pens that look like felt-tip pens. When you take off the cap, they emit smells of pineapple, tar, or fish. Healthy people can recognize and distinguish the different scents. In a further test, the doctor checks which concentration of the fragrance can still be perceived. In some cases, further examinations with imaging procedures or neurological tests may be necessary.

Therapies with Varying Success

For the common cold, as well as for a Corona infection, there is a first aid tool: patience. Usually, the olfactory mucosa manages to regenerate, although this can take several weeks or even months in the case of a Corona disease. Those who want to speed up the regeneration can train their nose with various essential oils in the morning and evening – this also needs to be maintained for several months. If the access of the scent molecules to the olfactory mucosa in the nose is permanently impaired or interrupted, such as by inflammation, deviated septum, or polyps, surgical intervention can be helpful. In the case of chronic inflammations, especially of the sinuses, as well as allergic mucosal swellings, medications such as antibiotics, antiallergics, or cortisone are used—often very successfully. Whether vitamins or growth hormones can accelerate the regeneration of dead olfactory cells by stem cells is controversial. The success of larger surgical interventions is also disputed. While changes in the mucosa are removed, the subsequent scarring of the tissue often blocks access to the olfactory mucosa again.

Unfortunately, all therapeutic approaches to healing olfactory cell damage caused by viruses, if the stem cells are also affected, have been completely unsuccessful to date. The same applies to accidents in which the ethmoid bone of the skull is affected and the small tubes of the ethmoid bone that allow the nerve fibers of the olfactory cells access to the brain are destroyed.

Smell Exercises for Greater Quality of Life

The most common cause of anosmia, old age, can not only be prevented with regular training, it can also slow down the loss of the sense of smell. The positive side effect is: those who practice smelling also train their brain.

From about the age of 60 on, the sense of smell gradually decreases, so it is important to start smell exercises early. Ideally, take a few minutes two or three times a day to deliberately smell some fragrant objects. This can include various types of fruit or vegetables, deodorants, creams or perfumes, as well as different juices, wines, or high-quality essential oils without additives. At the beginning, oils that can be easily distinguished are most suitable: rose, lemon, clove, eucalyptus, peppermint, or even spices like thyme and rosemary. The scents can be deeply inhaled with both nostrils or the nostrils can be trained individually—that is, covering the one and inhaling with the other, repeating this process a few times before switching nostrils.

It is important to engage with the objects with closed eyes and to ask yourself: What is this scent called? Do I readily recognize it? Where did I first encounter it? And what memories and emotions do I associate with the scent? Scientific studies at the University Hospital in Dresden have shown that, after half a year of training, over 30% of people have achieved an improvement in their sense of smell. In addition, the decline in olfactory ability in old age was delayed by a few years. The regained or improved sense of smell also had a positive effect on people's quality of life and mood. These studies also show that, in all people, a regular, conscious sniffing of fragrant objects improves the ability to perceive and identify smells.

Smell Training as Brain Jogging

Biologically speaking, every person has a set of 400 different types of olfactory receptors and about 20 million olfactory cells in the nose. The degree to which someone can successfully smell, identify and distinguish scents depends mainly on how intensively they train their sense of smell. A perfumer has the same equipment as we all do, but they practice "smelling" one to two hours every day. During this time, they focus all their attention on the respective scent. We, the average noses, can do this too. And the sooner we start, the better.

We can even pass on good olfactory abilities to our children. We should guide them to smell flowers, or the food that they are about to eat or drink, and maybe even consciously sniff at their fellow human beings when they hug them. Or to "smell around" a room when they enter it, even before looking around.

With conscious smelling, we can also give children and teenagers the opportunity to distinguish natural products from synthetic imitations for example, the original mango scent from the synthetic, as well as fresh orange juice from artificial or the scent of the vanilla pod from mere vanilla sugar. This allows teenagers and adults to identify, and perhaps even prefer, regional products, strengthening sustainability. Given all of this, there should actually be smelling lessons in school.

An additional effect that scientists have found in their studies during smell training was the influence on the brain. Those who welcome the emotions and memories evoked by the individual smells during the scent exercises activate significant parts of their brain. This has been so well established that, today, it is even referred to as "brain training" or "brain jogging". Similar positive results are also found in people with a reduced sense of smell (hyposmia).

A six-month course of training was able to help them largely restore their sense of smell. It is suspected that this is because olfactory cells grow back more strongly and the brain is again able to process the incoming signals correctly. Here, too, significant quality of life returns.

28

The Future: eNoses in Medicine

> It sounds like utopia, but it's no longer magic: In the future, there will not only be natural, but also electronic noses, abbreviated as eNoses. The eNose can be helpful in many areas of everyday life to reliably detect fragrances. For example, in medicine. Even today, we know that sick people smell different from healthy ones. Soon, the eNose could provide for quick diagnoses and enable appropriate therapies at an early stage.

An electronic nose is a technical system for odor measurement and digitization of the sense of smell. This makes it possible to develop cost-effective and practical olfactory systems that can be used in industry, food analysis, quality monitoring, environmental protection, and, especially, in medicine. With the help of microelectronic sensors, the eNose detects gaseous compounds in the air and converts them into a specific electrical pattern, depending on the signal strength of the individual sensors. The CO_2 detector, the exhaust gas detector in cars, is actually a simple example of an eNose.

Unlike the human and animal noses, the eNose does not absorb individual molecules through a specific odor receptor, but rather has various gas sensors—usually 5–40 pieces—and thus covers a wide range of gaseous chemical compounds. This provides an image of the composition of the air sample. However, the sensor cannot distinguish between odorless and odorous gases, nor does it perform any weighting or evaluation. It lacks the analysis that only a human brain can currently provide. However, new developments in AI have enabled the first prototypes that also include the ability to analyze.

The operation of the sensors varies. Some utilize a mass effect that uses the recognition of complex patterns and is particularly suitable for higher molecular weight fragrances. Others work with electrically conductive polymer compounds for polar (charged) fragrances. Most of the scents that our nose can smell are low molecular weight fragrances. The eNose detects such substances with semiconducting metal oxides. And then there are indeed eNoses that use a natural odor receptor from the nose of humans and animals and only use electronic components for the downstream analysis systems.

Until now, we have relied on natural scent detectors, especially the keen noses of dogs, to sniff out drugs at airports and some human diseases. Diseases change the body odor of humans. Sweat, breath and urine can all smell different when a person is suffering from diabetes, cancer, Parkinson's, or a bacterial infection. There is now much scientific evidence for this. Trained dogs can perceive the smell of diseases, but they require long and expensive special training to be suitable as medical sniffer dogs. Moreover, they can only be used for a limited time. They tire of the task relatively quickly, and they can only use their abilities at the highest level for a few years.

Thus, this endeavor would be more effective with electronic noses. It is conceivable that we might develop an eNose that analyses the patient's scent upon their entry into the doctor's office and sends the doctor hints about possible diseases to their computer. Or there could be breath test devices, similar to those for alcohol, that detect diseases or drugs. Or an eNose might be installed as a chip in the toilet, detecting tumor diseases in the bladder and intestinal area early and notifying the user on their mobile phone. Good advice from the chip-doc: visit a doctor!

Detecting Diseases and Germs Earlier

Experts agree that the eNose will bring enormous progress in the diagnosis and, especially, the early detection of diseases. The study "Smellodi," conducted at the University of Dresden by the research group of Prof. Thomas Hummel, is expected to provide initial insights. "Smellodi" stands for "Smart Electronic Olfaction for Body Odor Diagnostics," which can be translated as "Intelligent Electronic Odor Sensor for Body Odor Diagnostics." The project is now being continued by Prof. Ilona Croy at the University of Jena, a European center for the development of digital noses. It is being funded by the EU with millions of euros and has made Jena a hotspot for olfactory research. Sensors are being developed that can detect diseases when body odor changes.

The participants of the "Smellodi" study, who suffered from diseases such as Covid, Parkinson's, and the plain old common cold, provided valuable odor samples in the form of worn T-shirts. These were then given to both "test sniffers" and an eNose to smell. The eNose consisted of 4 chips with an integrated AI module, which can now already distinguish over 10 odors. However, so far, the trained human sniffing noses still come out ahead, recognizing Parkinson's, for example, with 78% accuracy. Further training of the eNose with AI data will likely soon reverse the situation.

With the development of such a system, expansion into many other areas, such as the food industry, would also be conceivable. In supermarkets or even in one's own refrigerator, eNoses could indicate spoiled food and provide important services to consumers. This would open up a mass market in the food sector, in addition to the development of highly specialized diagnostic devices in medicine. However, this requires not only the Sensors of the eNose to be developed, but first and foremost, a thorough exploration of the different smells depending on the disease. In the air exhaled by humans alone, more than 500 different fragrances can be found, some of which occur in higher or lower concentrations in certain diseases. In addition, there are previously unknown, new scents that have not yet been integrated into this spectrum, such as the odors of certain cancers. A wide field of research still awaits here. Furthermore, the composition of the scents in the breath also depends on many other parameters, such as time of day, temperature, food, fatigue, stress, etc. This complicates the establishment of a "healthy control scent."

Another important use of electronic noses in the medical field is the detection of potentially dangerous bacteria or germs in infectious diseases, such as the hospital germ MRSA. The responsible parties here are staphylococci, which are resistant to antibiotics and also occur in many bacterial urinary tract infections. Microbiologists have long known that different bacterial strains emit very specific fragrances. These allow scientists to recognize the bacteria by smell, even without laboratory analysis. Modern electronic noses can now take on such tasks faster and much more reliably than their human counterparts. In the laboratory, the eNose also performs an important task in quality control: it is capable of detecting fragrances that occur, for example, during the contamination of chemical processes. And it can help to avoid the use of contaminated products accordingly.

Not only doctors, but also police officers could benefit from the eNose. Its use in the drug scene is often tedious. Instead of having to rely on their own noses during checks, a test device—similar to the breathalyser—could make their work easier. The scent of cannabis contains, in addition to the typical cannabinoids, certain other fragrances in higher concentration, especially terpenes such as linalool, limonene, myrcene, pinene, or the humulenes from hops, as well as various flavonoids, which are often found in different plants, such as carvone from caraway. Sensors have already been specifically developed for this purpose that can be used permanently as an electronic nose.

29

Truffles, Tea, and Bugs: Searching for Species with the eNose

> Where are the truffles hiding? Is the tea pure or diluted with cheap products? And is the meat still edible or spoiled? With eNoses, these questions are quickly and safely answered. They react to specific scent patterns that they learn from biological noses. Particularly interesting for tourists: sensors that trigger an alarm at the smell of bedbugs.

The human nose consists of about 10 million sensory cells with approximately 400 different odor receptors. These receptors are specific to certain scents and generate a characteristic excitation pattern when activated by a mixture of fragrances. The brain then assigns this complex pattern to a specific smell that we have learned. This biological nose was the model for the development of an eNose at the Karlsruhe Institute of Technology (KIT), one of the leading scientific institutions for the development of electronic noses. A few years ago, researchers at KIT succeeded, for the first time, in developing an electronic nose for industrial and general applications, called "Kamina," whose name stands for "Karlsruhe Micro Nose." Such an eNose should be suitable for everyday use and be able to detect dangers such as cable fires or spoiled food better, faster, and more durably than our human or animal noses.

"Kamina" reacts, with the help of nanofibers, to complex gas mixtures, subsequently forming a complex signal pattern. Sensors recognize the corresponding scent from this. The eNose, with all of its evaluation technology for scent detection, is actually only a few centimeters in size. It is ideally suited as a measurement system, e.g., for cannabis, as it has high sensitivity and specificity, even for more complex scent mixtures. Above all, it is also robust and suitable for everyday use.

With a similar electronic nose developed at KIT, it was possible, for the first time two years ago, to distinguish different plants by scent. For example, there are many different types of mint that look similar, but all have a different scent. Here, even botanists often fail to identify the plant solely by the leaves. Only its scent reveals which type it is. At KIT, a chip was developed that has 12 specific sensors with 2 electrodes each and a quartz crystal on its surface, to which the fragrances can attach. This changes the so-called resonance frequency. From this data, a specific pattern of the respective scent is created, comparable to a fingerprint. With an attachment on the phone, the signals read from the chip can then be transferred to an app and analyzed there.

One interesting field of work for such chips is the verification of the purity of certain products, e.g., tea. Tea varieties differ significantly in price, and it would be interesting for consumers to know whether these prices are justified or if the tea has possibly been diluted with cheap varieties. The next plan of the researchers at KIT is to develop an eNose that can recognize and locate truffles. This is also a very rewarding goal, as truffles are expensive and grow invisibly underground. For a long time, efforts have been made to develop a method for finding them reliably. Two decades ago, there was already a very large and expensive "eNose" in the form of a miniaturized gas chromatograph. I once had the opportunity to use this in the Périgord. It was a device that looked like a large lawnmower, roaming through the deciduous forests of Tuscany. Whenever the device detected the scent of a truffle, an acoustic signal would sound, and one would then dig specifically at that spot. The yield was astonishingly high. Such a device could also be further developed for locating rare, endangered plants. This would be extremely significant for environmental policy in establishing appropriate protective measures.

NOSI Warns of Rotten Fruit—and Bedbugs

The Startup of the Year 2024 in Austria was the electronic sniffer by NOSI (Network for Olfactory System Intelligence). NOSI was founded as a spin-off of the work of the research institute AIT (Austrian Institute of Technology) in Vienna. NOSI has developed a digital nose that teaches machines to smell. The eNose relies on the interaction of chemical sensors that generate specific scent patterns. The machine learns from humans. Once it has been trained in the pattern of "rotten bananas" and has stored that information, it can recognize it at any time.

The highlight of NOSI: The chemical sensors are special polymers. They change their electrical conductivity when molecules of a substance, which is to be recognized by its smell, hit their surface. Just like in the human olfactory system, the chemical sensors are now calibrated to react like olfactory receptors to a specific scent facet—and only that facet. Because 16 sensors are mounted on a circuit board, all reacting slightly differently to the same scent molecule, different scent patterns can be measured.

The first project for NOSI was the detection of bedbugs. Bedbugs pose a real problem in the tourism industry—effective help against these annoying parasites is therefore very welcome. For this purpose, a sensor unit, located in a small plastic container and powered by a mini photovoltaic module, is attached to the bed. As soon as the smell of bedbugs or their excrement is detected, the system sounds an alarm. There are also considerations for further applications in the food industry or in the monitoring of technical systems, e.g., refrigeration shelves. However, it may take some time before a concrete product is realized.

At the University of Jena, a leading center for the development of digital noses, work is already underway. The chemical and food industries and environmental technology are possible areas of application. The smart home, the intelligent home of the future, could also benefit if the eNose automatically monitors the quality of stored food in the refrigerator. No more forgotten sausage in the back corner, no more spoiled fruit or moldy leftovers. "The market prospects for such a technology are enormous and range from devices for the mass market to highly specialized diagnostic devices," says Dr. Alexander Croy from the Physical Chemistry department at the University of Jena.

In Bavaria's largest refinery in Neustadt an der Donau, electronic noses have also become part of the safety concept. They react to odors that can be harmless, like crude oil, but also detect dangerous compounds like methane. The electronic noses are active around the clock and immediately report when certain hazardous substances are detected.

30

Nanoprostheses and Natural Replacement Noses

> Anyone who has lost their sense of smell due to Corona or another viral infection would give a lot to be able to smell and taste again. To date, such a loss is incurable. In the future, implanted electronic nano-noses or even models that work with human olfactory receptors could help these people.

During a Corona infection, about 75% of patients temporarily or even permanently lose their sense of smell. They become anosmic. They can no longer enjoy food because, with the loss of smell, taste is also lost. They do not perceive the scent of flowers, any season of the year or their favorite perfume. What they took for granted has suddenly disappeared. And, as is often the case, one only truly feels the loss of something once it is gone. In a recent study, American researchers found that there are at least 139 diseases associated with a disturbance of the sense of smell.

Anosmics particularly suffer from not being able to perceive their own body odor. They often wash several times a day to avoid smelling bad. They also cannot smell their partner's scent, which often affects the relationship, most especially, of course, the sexual aspect. It is not surprising that about 50% of affected Corona patients suffered from mild to severe depression. For these people, regaining their sense of smell would be of great importance. Fortunately, for many Corona patients, the condition proved to be temporary, as the olfactory cells in the nose renew themselves every 1–2 months from underlying stem cells. However, if the stem cells are also destroyed by the virus, there is currently no medical way to replace them and restore the sense of smell. These people remain anosmic forever. For hearing loss, there are hearing aids or cochlear implants; for vision loss, glasses or retinal implants;

for smell, there is currently nothing. Here, the great hope rests on the development of an eNose that can help in the form of an implant.

First prototypes already exist: A small chip is supposed to bring back the sense of smell. Researchers in Virginia, USA, are working on such a nanoprosthesis. The idea behind it: A sensor detects odors and converts this information into electrical signals for an external transmitter. This transmitter sends the signals to an implanted simulator, whose electrical stimuli can then trigger the sensation of smell in the brain. So far, only the external elements of the neuroprosthesis have been developed for this project. It contains sensors that are also used at the Karlsruhe Institute of Technology (KIT) to produce their eNose "Kamina," which we introduced in the last chapter. In addition to these sensors, there is a transmitter connected to a light-emitting diode.

In initial animal experiments employing mice, the generated signals directly stimulated the olfactory bulb. From there, they are transmitted to other brain areas and can, in turn, trigger reactions. Researchers suspect that mappings for different odors are created in the process. However, such e-noses do not yet exist for humans.

E-Noses from Natural Odor Receptors

A few years ago, my team at Ruhr University Bochum had the idea to use human odor receptors directly as an electronic nose. For this, the receptor proteins must be isolated from olfactory cells or from other cells that have been genetically modified to produce the desired odor receptors in large quantities. Human kidney cells are suitable for this, as well as frog eggs or yeast cells. This is now possible and feasible thanks to new developments in molecular biology and genetic technology. The problem with the odor receptors of mammals, including humans, is that the receptor proteins are very sensitive to certain environmental factors, such as temperature, humidity, and physiological solution.

Additionally, each individual receptor must activate a complex signal amplification system to generate a signal. This means that the activation of the receptor alone does not produce a signal sufficient for an electronic nose. The other molecules necessary for signal amplification must also be considered. In human kidney cells, these are automatically present and can be used to subsequently read the signals in other ways—for example, optically. The same applies to yeast cells into which the human receptor has been introduced. As a prototype, we used cell culture plates as a chip, with kidney cells into which we had introduced 10 different human odor receptors. They responded to different fruity scents, such as limonene, octanal, citronellal, and terpineol. We linked the activation of the receptor to a green color signal. After adding orange scent, a characteristic color pattern indeed emerged. Wonderful! We had hoped for and expected this, yet such an event is a highlight in laboratory life. If it becomes possible in the future to place all 400 human receptors on such a plate, this chip would be comparable to a human nose.

A similar project, also using original odor receptors, is currently being developed by Professor Klemens Störtkuhl's laboratory, also at RUB. In this project, odor receptors of the fruit fly (Drosophila) are used. Flies have about 70 different odor receptors, primarily for food and danger scents, that react very sensitively and specifically to certain odorants. Insect receptors are fundamentally different from vertebrate odor receptors. They have the advantage that they do not require additional amplification molecules. Their activation alone generates a sufficient electrical signal.

Such odor receptors can also be produced in genetically modified frog eggs. Small membrane patches from the eggs, in which the receptors are located, can be punched out and applied to the tip of electrodes. These are then used as an eNose. Initial successful experiments with, for example, CO_2 sensors have already been conducted. However, using a natural receptor has the disadvantage that the lifespan is limited to a few hours. Advantages, however, include high sensitivity and specificity, as well as the fact that they also respond quantitatively to odor concentrations. This could make it possible in the future to detect very specific odor molecules that occur only in low concentrations in certain diseases, such as cancer or neurodegenerative diseases, in exhaled air or sweat, at an early stage.

However, the broad application will likely belong to the electronic noses developed by engineers. They are more robust and less sensitive to external influences such as temperature, humidity, and other environmental factors. The future will feature e-noses being found everywhere in our lives in the next two decades—from supermarkets to doctor's offices to smart homes. They will warn us of all kinds of dangers and help us detect diseases in a timely manner.

GPSR Compliance
The European Union's (EU) General Product Safety Regulation (GPSR) is a set of rules that requires consumer products to be safe and our obligations to ensure this.

If you have any concerns about our products, you can contact us on

ProductSafety@springernature.com

In case Publisher is established outside the EU, the EU authorized representative is:

Springer Nature Customer Service Center GmbH
Europaplatz 3
69115 Heidelberg, Germany

www.ingramcontent.com/pod-product-compliance
Lightning Source LLC
LaVergne TN
LVHW010344260326
834688LV00036B/867